Effective Angular

Develop applications of any size by effectively using Angular
with Nx, RxJS, NgRx, and Cypress

Roberto Heckers

Effective Angular

Group Product Manager: Kaustubh Manglurkar

Publishing Product Manager: Chayan Majumdar

Book Project Managers: Shagun Saini and Arul Viveaun S

Senior Editor: Anuradha Joglekar

Technical Editor: K Bimala Singha

Copy Editor: Safis Editing

Proofreader: Anuradha Joglekar

Indexer: Rekha Nair

Production Designer: Prashant Ghare

DevRel Marketing Coordinators: Mouli Banerjee and Nivedita Singh

Publication date: August 2024

Production reference: 2160824

Published by Packt Publishing Ltd.

Grosvenor House

11 St Paul's Square

Birmingham

B3 1RB, UK.

ISBN 978-1-80512-553-2

www.packtpub.com

Contributors

About the author

Roberto Heckers is an Angular specialist who has been using the framework since the AngularJS days. He has extensive experience under his belt with the Angular framework. His journey started as a self-taught software engineer building applications with AngularJS. In 2019, he was a front-end team lead for the first time, building an extensive real-estate application and UI library for Spacewell. Since then, he has been focused on building enterprise applications with complex architecture and scalability needs. Roberto now works as a freelance Angular engineer for some of the biggest companies in the Netherlands, such as Vattenfall and Marel, building applications used by millions of users.

About the reviewers

Sarath Raj, a seasoned professional with over 15 years of experience, specializes in software design, development, and maintenance across diverse sectors such as business-to-business, trading platforms, and customs border management. With proficiency in Reactive microservice architecture, cloud-native apps, DevOps, and mobile apps, he is recognized for his innovative product architectures and adeptness in concurrent project execution. He is deeply passionate about application security, Rust, and Solana blockchain, and his expertise spans Java, Angular, Scala, Kotlin, distributed systems (Akka), the cloud (Azure and Google), and data analytics (Azure and Google).

Rubén Peregrina is a lead frontend engineer hailing from Barcelona, Spain. With over seven years of experience, he is an expert in Angular and has worked in various sectors, such as e-commerce, marketing, healthcare, and fintech. Rubén started his career as a full stack developer, working with AngularJS and Java. He then went on to explore other technologies, such as Vue.js and Flutter. In recent years, he has focused primarily on Angular, working on international projects. Aside from his work, Rubén is an active contributor to different projects on GitHub and has even started his own blog. Through his blog, he shares his expertise in frontend development, Angular, TypeScript, and JavaScript.

Table of Contents

2

Powerful Angular Features   35

3

Enhancing Your Applications with Directives, Pipes, and Animations 69

4

Part 2: Handling Application State and Writing Cleaner, More Scalable Code

5

6

7

Mastering Reactive Programming in Angular   185

8

Handling Application State with Grace   223

Part 3: Getting Ready for Production with Automated Tests, Performance, Security, and Accessibility

9

Enhancing the Performance and Security of Angular Applications   261

10

Internationalization, Localization, and Accessibility of Angular Applications   283

11

Testing Angular Applications   305

12

Deploying Angular Applications   339

Preface

Welcome to *Effective Angular: Develop applications of any size by effectively using Angular with Nx, RxJS, NgRx, and Cypress*. This book is your comprehensive guide to mastering Angular and leveraging its powerful ecosystem to build scalable and professional web applications. Whether you are a seasoned Angular developer or just starting your journey, this book will equip you with the knowledge and skills needed to build scalable, enterprise-grade applications with confidence.

Angular is a powerful frontend framework developed and maintained by Google. Since its inception, it has become one of the most popular frameworks for building dynamic web applications. Angular's component-based architecture, coupled with its strong TypeScript foundation, provides developers with a robust and scalable platform to craft modern applications.

In this book, we will delve into the intricacies of Angular, guiding you through its features and best practices. This book is not just about writing code; it's about understanding the underlying principles that make Angular such a versatile tool to build applications of any size. From setting up scalable environments with Nx to mastering reactive programming with RxJS and NgRx, each chapter is crafted to provide you with practical insights and real-world examples that you can apply directly to your projects.

You will learn how to design scalable frontend architectures and set up Angular monorepos using Nx. Dive deep into Angular's most powerful features, including dependency injection, routing, Signals, directives, pipes, animations, and standalone components. You will discover how to manage application state using NgRx and understand reactive programming with RxJS and Angular signals, ensuring your applications are both efficient and scalable.

This book will guide you through the complexities of the Angular ecosystem, empowering you to build professional-grade applications with ease.

Who this book is for

This book is tailored for front-end engineers looking to level up their Angular skills. Whether you're a seasoned developer or just starting out, this book will guide you through the complexities of the Angular ecosystem, empowering you to build professional-grade applications with ease. You will learn how to build applications of any scale by utilizing the powerful features within the Angular framework. Hands-on examples will teach you about the newest Angular features such as standalone components, Signals, and control flow. You will learn about frontend architecture, building Angular monorepo applications using Nx, reactive programming with RxJS, and managing application state with NgRx. By the end of this book, you will know all Angular's features to use it effectively.

What this book covers

Chapter 1, Scalable Front-End Architecture for Angular Applications, provides you with knowledge about frontend architectures, enabling you to make the right decisions when setting up your workspace and developing scalable frontend systems. In this chapter, you will also create your own Nx monorepo, ready to handle hundreds of Angular applications and libraries without breaking a sweat.

Chapter 2, Powerful Angular Features, brings you up to speed with the latest developments and covers some of the most powerful features of the Angular framework. You will learn about new concepts such as standalone components and signals. Additionally, this chapter will provide a deep dive into the Angular router, component communication, and dependency injection.

Chapter 3, Enhancing Your Applications with Directives, Pipes, and Animations, shows you how to write more reusable code by utilizing directives and pipes. You will learn about best practices, common pitfalls, and design choices when developing custom pipes and directives, or when using the built-in options of the Angular framework. Furthermore, this chapter teaches you about creating and reusing UI animations within the Angular framework to spice up your web applications.

Chapter 4, Building Forms Like a Pro, provides a step-by-step guide to developing template-driven, reactive, and dynamic forms. You will learn how to validate your input fields for each approach and how to handle error messages. By the end of this chapter, you will have built a template-driven form, converted it into a reactive form, and finally, made the same form using a dynamic form approach.

Chapter 5, Creating Dynamic Angular Components, focuses on developing reusable components for complex design needs. You'll learn about content projection, projection slots, template variables and references, dynamically creating components, lazy-loading non-routed components, and developing your own widget component, with the ability to load and mount components on demand based on user input.

Chapter 6, Applying Code Conventions and Design Patterns in Angular, discusses how to incorporate structure and patterns into your Angular applications. From code conventions and best practices to commonly used design patterns such as the facade, decorator, and factory pattern, this chapter will give you the knowledge and structure needed to develop scalable and robust Angular applications. We will end this chapter by developing a generic type-safe HTTP service with a model adapter that automatically converts your DTOs to view models, and vice versa.

Chapter 7, Mastering Reactive Programming in Angular, emphasizes the advantages of reactive programming and shows you how to implement it in your Angular applications using RxJS and signals. You will learn to handle asynchronous data streams using RxJS and synchronous data with signals. We will explore how RxJS and signals can be combined and distinguish when to use RxJS and when to use signals to handle reactivity within your Angular applications.

Chapter 8, Handling Application State with Grace, teaches you everything you need to know about application state and how to handle it gracefully within your Angular applications. We will create a facade service to connect your component layer with your application state and make three different implementations to handle the global application state. We will start by creating a custom application state solution using RxJS. Then, we will convert it to a custom solution using signals. Lastly, we look at the shortcomings of RxJS and signals when handling the global application state of complex applications and implement NgRx to mitigate these shortcomings.

Chapter 9, Enhancing the Performance and Security of Angular Applications, helps you develop more performant Angular applications by providing a detailed explanation of the Angular change detection mechanism and showcasing other factors that impact application performance, showing you how to reduce their impact. Additionally, this chapter teaches you how to develop secure Angular applications by exploring possible attack surfaces for malicious actors and how to mitigate them.

Chapter 10, Internationalization, Localization, and Accessibility of Angular Applications, enables you to develop Angular applications accessible to people with different impediments, speaking different languages, or who are used to various formats. You will implement the Transloco library for i18n and i10n and learn about web accessibility guidelines.

Chapter 11, Testing Angular Applications, provides you with hands-on experience in writing automated tests for Angular applications. You will learn about the different types of tests used in frontend applications and how to write and run unit tests using Jest and end-to-end tests using the Cypress test framework.

Chapter 12, Deploying Angular Applications, outlines the steps for building and deploying Angular applications located in an Nx monorepo. You will learn how to analyze your bundle sizes, run linting, and build commands for one or more projects in your Nx monorepo and how to automate your deployment process.

To get the most out of this book

To get the most out of this book, you should have experience building small to mid-sized frontend applications with modern frontend frameworks. You should also know at least the basics of the Angular framework.

Software/hardware covered in the book	Operating system requirements
Angular 17.1	Windows, macOS, or Linux
TypeScript 5.3.3	
Nx 18.0.3	
Cypress 13.0.0	
Jest 29.4.1	

If you are using the digital version of this book, we advise you to type the code yourself or access the code from the book's GitHub repository (a link is available in the next section). Doing so will help you avoid any potential errors related to the copying and pasting of code.

Download the example code files

You can download the example code files for this book from GitHub at `https://github.com/PacktPublishing/Effective-Angular`. If there's an update to the code, it will be updated in the GitHub repository.

We also have other code bundles from our rich catalog of books and videos available at `https://github.com/PacktPublishing/`. Check them out!

Conventions used

There are a number of text conventions used throughout this book.

`Code in text`: Indicates code words in text, database table names, folder names, filenames, file extensions, pathnames, dummy URLs, user input, and Twitter handles. Here is an example: "I will only add the following configurations in the `jest.preset.js` in the root of the Nx monorepo."

A block of code is set as follows:

```
const mockTranslationService = {
  translocoService: { translate: jest.fn() },
  translationsLoaded: signal(false) as WritableSignal<boolean>,
};
```

When we wish to draw your attention to a particular part of a code block, the relevant lines or items are set in bold:

```
TestBed.overrideComponent(AppComponent, {
  add: {
    imports: [StubNavbarComponent],
  },
  remove: {
    imports: [NavbarComponent],
  },
});
```

Any command-line input or output is written as follows:

```
npx nx run <project-name>:test
```

Bold: Indicates a new term, an important word, or words that you see onscreen. For instance, words in menus or dialog boxes appear in **bold**. Here is an example: "In the browser, you'll see **Expenses Overview**, just like before."

> **Tips or important notes**
> Appear like this.

Get in touch

Feedback from our readers is always welcome.

General feedback: If you have questions about any aspect of this book, email us at customercare@packtpub.com and mention the book title in the subject of your message.

Errata: Although we have taken every care to ensure the accuracy of our content, mistakes do happen. If you have found a mistake in this book, we would be grateful if you would report this to us. Please visit www.packtpub.com/support/errata and fill in the form.

Piracy: If you come across any illegal copies of our works in any form on the internet, we would be grateful if you would provide us with the location address or website name. Please contact us at copyright@packt.com with a link to the material.

If you are interested in becoming an author: If there is a topic that you have expertise in and you are interested in either writing or contributing to a book, please visit authors.packtpub.com.

Share Your Thoughts

Once you've read *Effective Angular*, we'd love to hear your thoughts! Scan the QR code below to go straight to the Amazon review page for this book and share your feedback.

https://packt.link/r/1-805-12553-2

Your review is important to us and the tech community and will help us make sure we're delivering excellent quality content.

Download a free PDF copy of this book

Thanks for purchasing this book!

Do you like to read on the go but are unable to carry your print books everywhere?

Is your eBook purchase not compatible with the device of your choice?

Don't worry, now with every Packt book you get a DRM-free PDF version of that book at no cost.

Read anywhere, any place, on any device. Search, copy, and paste code from your favorite technical books directly into your application.

The perks don't stop there, you can get exclusive access to discounts, newsletters, and great free content in your inbox daily

Follow these simple steps to get the benefits:

1. Scan the QR code or visit the link below

https://packt.link/free-ebook/978-1-80512-553-2

2. Submit your proof of purchase
3. That's it! We'll send your free PDF and other benefits to your email directly

Part 1:
Angular Basics and Setting Up Scalable Nx Workspaces

In this part, you'll learn about frontend architecture and what to consider when creating your workspace. Additionally, you'll set up your own Nx monorepo, which is ready to handle hundreds of Angular applications and libraries. After creating your Nx monorepo, you'll learn about the newest features in the Angular framework, such as signals and standalone components. When you're fully up to date with the latest developments of the Angular framework, you'll do a deep dive into the Angular router, component communication, and dependency injection. Then, you will get hands-on experience with Angular directives, pipes, and animations and learn about the best practices, their impact on performance, and common pitfalls when using directives, pipes, or animations. You will finish the first part by creating template-driven, reactive, and dynamic forms, understanding their differences when you use each method and how you can adequately validate them.

This part includes the following chapters:

- *Chapter 1, Scalable Front-End Architecture for Angular Applications*

- *Chapter 2, Powerful Angular Features*

- *Chapter 3, Enhancing Your Applications with Directives, Pipes, and Animations*

- *Chapter 4, Building Forms Like a Pro*

1

Scalable Front-End Architecture for Angular Applications

Angular is a powerful and extensive framework for building web applications. According to the 2023 Stack Overflow developer survey, it is the fourth most used web technology after ReactJS, NodeJS, and jQuery among professional developers. Due to the structure and tools it provides, Angular is often chosen when building large web applications or enterprise solutions comprised of several applications and libraries.

This book will guide you through the process of effectively using the Angular framework to develop and test applications of any size. You'll start by learning about front-end architecture and setting up a scalable workspace with Nx that's ready for hundreds of Angular applications. Next, you'll explore the most powerful and newest features within the Angular framework. You'll learn about reactive programming and state management using RxJS, Signals, and NgRx, and will be able to test Angular applications with Jest and Cypress. Upon completing this book, you'll be able to effectively use the Angular framework and develop scalable, enterprise-ready Angular applications, utilizing all the tools Angular has to offer while implementing best practices and sound design patterns.

In this chapter, you will create your Angular workspace using Nx. You'll start by learning what we mean by scalable front-end architecture and why it is essential to think about your architecture before you start writing code. You will also learn about different patterns in front-end architecture and what to consider when building enterprise-ready solutions from scratch.

Lastly, you will explore and use Nx, the build tool that allows you to create scalable Angular monorepos. By the end of this chapter, you will understand crucial aspects of front-end architecture and have your own Nx monorepo for Angular, ready to handle hundreds of applications easily.

This chapter will cover the following main topics:

- Understanding scalable front-end applications
- Different approaches to scalable front-end architecture
- What is Nx and why should you use it?
- Setting up a scalable Angular workspace

Technical requirements

By the end of this chapter, we will create an Nx monorepo with an Angular application and library in it. To follow along, you'll need to install some tools. Note that we will only use freely available tools.

You will require the following:

- **Visual Studio Code (VS Code)** as your **integrated development environment (IDE)**
- Chrome web browser
- Angular 17.1 or higher
- NodeJS version v20.11.0 or higher
- TypeScript version 5.3.3 or higher
- Nx version v18.0.7 or higher

Throughout this book, we will use Angular 17.1, NodeJS 20.11.0, TypeScript 5.3.3, and Nx 18.0.7.

The GitHub repository for this book is available at `https://github.com/PacktPublishing/Effective-Angular`.

Understanding scalable front-end applications

Modern web applications are constantly getting bigger and more complex. Because of this, developing scalable front-end applications is more critical than ever. To create scalable front-end applications, we need to understand what scalability means in the context of a front-end application.

What is scalability?

The first thing that might come to mind when you hear the term **scalability** is handling more traffic. However, in the context of front-end applications, when talking about scalability, we mostly mean the scalability of the code base. Or, more concisely, the code is easy to extend, and modules or micro front-ends can be added to the software without much work. Components and libraries are reusable; the code is easy to maintain, test, and debug, even when the application grows. You can work with different teams on separate parts of the application, and onboarding new teams that write similar code is easy to achieve and enforce. The application has good performance and small bundle sizes. Compile and build times remain low, and deployment can be done swiftly to different staging environments if needed.

To achieve these feats within your front-end applications, you must create a good architecture, a set of tools, and rules everyone can adhere to. Your architecture will include elements such as your repository type, folder structure, architectural patterns, design patterns, a programming language, framework, and tools for building, testing, linting, and deploying your application. Making the right decisions for each part of your architecture helps with creating scalable applications that are easy to maintain and extend.

When making architectural decisions, you should aim to create a fast, loosely coupled, testable system. You want to avoid direct dependencies between different parts of your system so that you don't get stuck and need to refactor the entire application when the business introduces changes.

With this brief introduction to what a scalable front-end application is, let's understand the importance of a good front-end architecture.

Why is front-end architecture important?

Good architecture is essential to maintain a good workflow for the developers working on the software and makes it easy for new teams and team members to join. If developers spend much time refactoring or waiting for builds or tests to complete, they wander off and do less productive things.

With good architecture in place, the code base remains manageable, even when your applications grow. Without good architecture, the code becomes messy and hard to debug or extend. As time progresses, such problems will pile up, and developers will be wasting more time on bugs and refactoring than creating new features, especially in large enterprise solutions where business needs are constantly changing. Before you know it, you will have a web of dependencies, and adding simple things will become very time-consuming.

Suppose you're building an application for employee scheduling that includes a calendar component. You want to avoid tight coupling between the calendar and scheduling applications. When management lets you know the company is adding another application – let's say a project management tool that also includes a calendar – you don't want to redo the entire calendar component and scheduling application because the two are tightly coupled.

The following figure shows the development process with and without architecture. Without architecture, you start fast, but in the end, you are crawling. With architecture, you will have a consistent and predictable work pace:

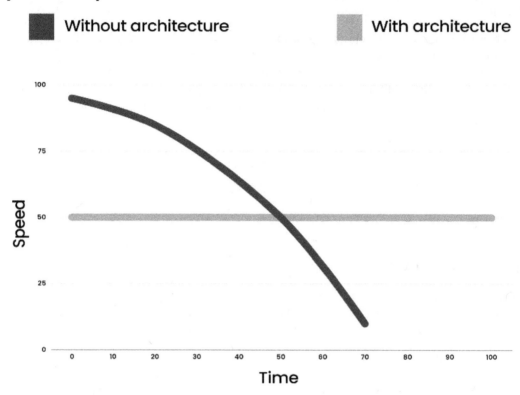

Figure 1.1: Development speed

Now that you understand what we mean by scalable front-end applications and why having good architecture is essential to achieve them, we will dive into some approaches to scaling front-end applications and learn about their advantages and trade-offs.

Different approaches to scaling front-end applications

When it comes to architecting software, it's essential to think of what you are building, what it can grow into, what the context is, and who will work on it. Depending on these parameters, you'll want to create an architecture that's flexible enough to grow and adapt without overengineering it, making things more complex and time-consuming than needed.

For example, if you're building a simple website for a small family-owned business, you don't need elaborate architecture and complex design patterns; this will make things more complex and time-consuming than they need to be. The needs for the website will probably stay mostly the same, and the code base will remain small and manageable. But when you're building enterprise software comprised of multiple applications, the needs and utility of those applications will change quite a bit, and you'll want to ensure that the software is set up for that.

In this section, you will learn about different architectural choices so that you can ensure you don't over or under-engineer your front-end applications. You will learn about different repository structures and architectural patterns that are commonly used within Angular applications. Without further ado, let's learn about the differences between mono and poly repositories.

Monorepo or polyrepo

Monorepo and **polyrepo** are two options for storing code within a source control application such as GitHub. If you create a new repository for each project or application, it's called a polyrepo or multi-repos structure. Meanwhile, when all applications are in one large repository, it's called a monorepo structure. For large companies, a monorepo can easily contain hundreds of applications.

What option should you pick when developing a large front-end environment? Let's start exploring the advantages and disadvantages of both solutions.

Advantages of monorepos

Some of the key advantages of monorepos are as follows:

- **Easy code sharing**: Monorepos make it easy to share code among different projects and teams. You have the code of all projects in one directory, so inspecting or reusing code from another project and sharing things in libraries is simple. This helps prevent duplicate code, gives you inspiration and guidance on solving similar issues, and allows you to quickly check if a bug originates in your code or the implementation of a library.

- **Uniformity among different applications**: Because all code resides in one repository, enforcing uniformity among different applications is easier. You can ensure that the same tooling is used, code conventions and best practices are applied, and all applications are tested similarly. Also, testing the entire system is easier with a monorepo.

- **No versioning conflicts**: In a monorepo, dependencies commonly share the same version among all applications. This ensures that no versioning conflicts occur and implementations are equal in all applications. Furthermore, it ensures that applications that are not actively developed still get regular updates for their dependencies.

- **Cross-project refactoring**: If you want to refactor something that occurs in many applications, you can do it in one go for all applications or efficiently run scripts to perform these actions.

- **Quick code movement and debugging**: Code can quickly be moved from one project to another, and when you encounter a bug within a library, you can fix the bug without holdups and continue your work.

- **Shared commit timeline**: Lastly, a monorepo has one shared commit timeline. This makes it safe to create atomic commits with changes in multiple applications. This happens because if anything goes wrong, you can always revert to a common state of all applications through the commit timeline.

Regardless of these advantages, monorepos have some shortcomings too. We'll look at them in detail in the following section.

Disadvantages of monorepos

Some of the disadvantages of monorepos are as follows:

- **Large folder structure**: A monorepo can be daunting because all code and libraries live in one large solution. If you have yet to work with a monorepo, this might take some time to get used to. Because it's so easy to use code from other projects, you need to be extra careful not to create unwanted dependencies between applications.

- **Complex package updates**: Updating packages and dependencies in a monorepo can be complicated because, often, all projects need to update at once.

- **Breaking changes**: Because there are no different versions, when you make changes in libraries, you can break other applications without noticing it.

- **Slow build times**: In a monorepo, building and testing can become time-consuming if not managed correctly.

- **Challenges in deployment**: Deploying applications modularly can be more challenging compared to polyrepos, where everything is separated and modular by nature.

Most disadvantages can be mitigated with the right tools. For monorepos comprised of Angular applications, you can use Nx, giving you everything to handle a monorepo without these disadvantages. In this book, we will work with a monorepo and Nx.

Now, let's move on to the advantages and disadvantages of polyrepos.

Advantages of polyrepos

The key advantages of polyrepos are as follows:

- **A higher level of isolation between applications**: The most apparent advantage of a polyrepo is higher levels of isolation between your applications. Each project has a repository and developers can do whatever they want within that repository without affecting other projects too much.

- **Flexibility with dependency management**: When using a polyrepo, each project can manage versions of its dependencies on its own. This gives the teams working on the projects more freedom when updating dependencies and offers more stability.

- **Individual tooling**: With a polyrepo, there is more flexibility for using different tools and programming languages.

- **Easy to manage**: A polyrepo is generally easier to manage, especially for smaller teams. There is little code in each repository compared to a monorepo, so there are fewer things you can affect with your changes.

- **Straightforward modular deployment**: Deploying your applications modularly is more straightforward.

- **In-line with a micro front-end architecture**: Lastly, if you're developing with a micro front-end architecture, using a polyrepo might feel more in tune with the rest of your architecture. However, micro front-ends can be achieved with both a monorepo and polyrepo structure.

Before you pick between a monorepo or polyrepo structure, you must also consider the drawbacks of polyrepos. These are described in the next section.

Disadvantages of polyrepos

Some common disadvantages of polyrepos are as follows:

- **Difficulties in code sharing**: First, because code resides in isolation, it is harder to share code between applications. This can result in different implementations for the same problem or duplicate code in various projects. Teams will often create their own solution instead of contributing to a library that solves shared problems.

- **Boundaries between projects and libraries**: When you depend on a library that resides in another repository, and you run into a bug, it will be more time-consuming to fix the issue. Often, you need to wait for other teams to fix the bug in the library and deploy a new version before you can continue.

- **Challenges when testing multiple repositories**: Testing applications can become more challenging. Predominantly when applications consist of different modules that reside in different repositories, testing the entire system can be difficult.

- **Difficulties in creating CI/CD pipelines and deployments**: Creating CI/CD pipelines and deployments of the whole system might become more challenging as you need multiple repositories to complete the tasks.

- **Dependency conflicts**: Lastly, you can run into dependency conflicts. Different applications need to work together in production and can depend on similar dependencies. You can encounter compatibility issues when these applications use different versions of the set dependency.

In this section, you learned about the advantages and disadvantages of storing your code in a monorepo or polyrepo. Next, you will learn about different architectural patterns that are commonly used with Angular applications.

Architectural patterns for Angular applications

Architectural patterns focus on how you structure your code and provide rules for abstracting business logic away from specific implementations. Architectural patterns are an extensive topic and there are entire books dedicated to it; so, it will be impossible to cover everything in this section. Still, we will briefly cover some of the most common architectural patterns that are used with Angular and Nx. In *Chapter 5*, we will dive into design patterns while focusing on code implementations instead of providing a high-level view of the system.

Now, let's dive into some architectural patterns and learn what they try to achieve, how they try to achieve it, and what their advantages and disadvantages are.

Common architectural patterns in Angular applications

Most architectural patterns try to accomplish the same at their core, only with little nuances and different terminologies. Regarding architectural patterns for Angular applications, they try to separate domain and business logic from implementations and the view. Doing so gives you a loosely coupled system that is easy to test, change, and expand without creating dependencies in the wrong places. In this section, we will cover the **Model-View-Controller** (**MVC**), hexagonal architecture, and layered architecture. These three patterns are some of the most commonly used architectural patterns for Angular applications. Other noteworthy patterns are the **Model-View View-Model** (**MVVM**), union architecture, and clean architecture.

Without further ado, let's learn about the MVC pattern and how it can be used within Angular applications.

MVC in Angular applications

MVC is one of the most commonly used architectural patterns in the world of software development. The pattern was first used in the back-end but is now also used for front-end applications, or at least something resembling the MVC pattern. The MVC pattern is often used for Angular applications because it fits well with the tools provided by the framework. As the name implies, MVC consists of three parts – the **model**, the **view**, and the **controller**:

- The model declares the data models and handles business and data-related logic.
- The view displays the current state of the model to the user.
- The controller acts as a bridge between the view and the model. The controller passes the user input to the model so that the model can perform actions and update accordingly, after which the controller returns the updated values to the view.

If we translate this pattern into an Angular application, the model would be a service handling the data and data models. The view would be the HTML template. Finally, the controller would be the TypeScript file behind the HTML file, commonly named the component class. A better implementation is considering the component class as part of the view and having an extra facade service (an additional abstraction layer creating a simple interface and communication layer between the view and business or state-related logic of your application; we will discuss the facade pattern in more detail in *Chapter 5*) between the model service and the view acting as the controller.

If you keep true to the MVC architecture's original implementation, the model directly gives the updated values to the view. To achieve this, in *Figure 1.2*, we would eliminate *steps 4* and *5* and would then go to the view instead of the controller. In Angular, this is sort of the case when using the component class as a controller, and it results in a tight coupling between the business logic and the component classes. Because of that, I prefer to add a separate facade service that acts as a controller to fully separate the component classes from the business and state layers of my application. By separating the components from the state and business logic, you end up with a loose coupling, making it easier to change implementations throughout your application:

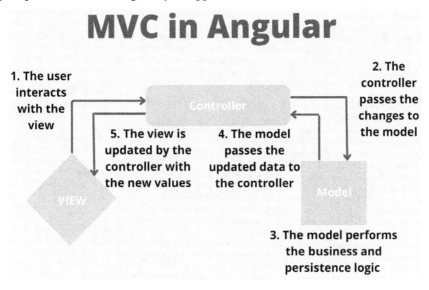

Figure 1.2: MVC pattern

Now that you know what the MVC pattern entails and how you can implement it within your Angular applications, let's learn about the next common architectural pattern: the Hexagonal architecture.

The hexagonal architecture pattern in Angular applications

Hexagonal architecture is relatively new compared to MVC and some other architectural patterns, such as layered architecture, MVVC, and MVP. The hexagonal architecture was introduced in 2005, and people have only started implementing it in Angular applications in the last couple of years. It gained popularity because **domain-driven development (DDD)** became a hot topic, and hexagonal architecture is very well suited to combine with DDD. The main principle of hexagonal architecture is to separate the core application logic away from the UI and data implementation through ports and adapters. Because of that, the architecture is also commonly referred to as the *ports and adapter architecture*. But I hear you thinking, what are ports and adapters?

In simple terms, **ports** are interfaces (or abstract classes) that separate your core logic from the UI and code implementations. These interfaces dictate how the UI and code implementations can communicate with your application core. The **adapters** are the UI and code implementations that connect with your application core through the ports. In hexagonal architecture, ports and adaptors come in two types, UI and data-related ports and adapters – in other terms, primary and secondary adapters and ports. This concept is illustrated in *Figure 1.3*:

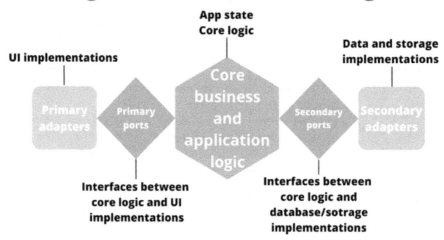

Figure 1.3: Hexagonal architecture pattern

When implementing hexagonal architecture, you will have a set of ports and adapters for each domain within your application. I like to use a facade service between the port interfaces and adapters for more abstraction between the UI, implementations, and application core. When using a facade service, the facade can be considered as the port itself. Just make sure the facade implements an interface so that there is a fixed set of rules for communication with the core and adapters.

Because ports define a fixed set of rules for communicating with your application core, you can easily change implementations when business requirements change. You can swap UI components without touching your business logic or data implementations, and you can change how you persist or fetch your data without touching your views or application core. The only thing you need to do is ensure your new implementation can connect to the same interface your ports are using to connect everything. This approach offers excellent flexibility and a loosely coupled system.

To clarify things, I want to go over *Figure 1.3* and translate it into an Angular application. We will go from left to right. On the far left, we have the primary adapters. Everything the user faces, or triggers, is considered a primary adapter: components, directives, resolvers, guards, and event listeners. One step to the right, we will find the primary ports. These regular TypeScript interfaces (or facade services) dictate how the UI layer communicates with the application core. Our application core is in the middle, where we access state management and define business and application logic in Angular services. On the right of the application core, we have our secondary ports. These ports dictate how the application code communicates with HTTP services, state management, in-memory persistence, event dispatchers, and other data or API-related logic. Like the primary ports, the secondary ports are regular TypeScript interfaces (or facade services). On the far right, we have our secondary adapters. The secondary adapters implement our HTTP services, local storage persistence, state management, and event dispatchers.

Now that you know what hexagonal architecture is and how you can implement it within your Angular applications, let's take a look at the third and final architectural pattern we will discuss: layered architecture.

The layered architecture pattern in Angular applications

As the name implies, the layered architecture pattern uses different layers to separate concerns. For each application section, you create a layer in the architecture that sits on top of another layer. When you implement the layered architecture pattern in your Angular applications, you should have at least three (main) layers: the core layer, the abstraction layer, and the presentation layer. Within these top-level layers, you can have additional sub-layers or sibling elements. If your application architecture needs more layers, you can add them as needed.

The most important thing with layered architecture is that each layer can only communicate with the layer above and below itself; the rest of the layers in the chain are off limits. Another essential feat is that events and actions flow upwards, and data flows downwards. The user triggers an event or performs an action in the presentation layer. This layer notifies the abstraction layer of the action and the corresponding changes. The abstraction layer sends these actions and changes to the core layer, where the business logic is performed and the data changes persist. When the core layer has performed the application logic and persisted the changes, the data will flow back from the core layer through the abstraction layer in the presentation layer, where the view is updated for the user:

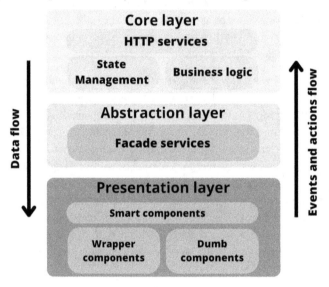

Figure 1.4: Layered architecture pattern

Throughout this book, we will use a layered architecture resembling what's shown in *Figure 1.4*, with the presentation layer containing dumb, wrapper, and smart components. Dumb components only have component inputs to receive data and outputs to alert the parent components something has changed. Wrapper components are also dumb, but they are used to group multiple components and provide a reusable layout or animation. Even though a wrapper container can wrap around dumb components, they are the same for the flow of data and separation of dependencies, which is why they are placed next to each other in the architecture design. On top of the dumb components, we have smart components. These components are, generally speaking, specific business use cases or pages, and they inject facade services and implement component logic and state.

The next main layer in our architecture is the abstraction layer, where we have facade services. These facade services are regular Angular services that implement the facade design pattern. These facade services provide additional abstraction and are used as a bridge between our smart components and the core layer of the application.

Our last major layer is the core layer, where the global state management, business and application logic, and HTTP services reside. Our HTTP services layer lies on top of the state management and business logic layers. We have separate services that do nothing but fetch data and pass it to our other core layers; the lower core layers never fetch data directly, so we have an additional abstraction layer and have better separation of concerns.

Now that you know about the layered, hexagonal, and MVC architectural patterns, let's move on and briefly learn about the advantages and disadvantages of each

Comparing the architectural patterns

All three patterns we discussed in the *Common architectural patterns in Angular applications* section separate the business logic from the implementations and presentation code. They all provide abstraction layers but have different approaches to creating these layers. First, the MVC pattern might seem the simplest to implement, but you can have too many dependencies and implementations in your component classes if you don't add facade services, especially if you let the model part of MVC directly communicate with the view. Not only will this tightly couple your view with your business logic, but it can also trigger extensive DOM updates. Changing implementations can become hard, and unit testing needs a lot of mocking.

Next, we have hexagonal architecture. I like this architecture, but it introduces a lot of boilerplate code and can feel complex to implement, making it not the right fit if you have a lot of junior developers on your team. Nonetheless, the significant advantage of hexagonal architecture is that you can easily change implementations once everything has been set up. Unit testing the code also becomes straightforward because everything is separated into ports and adapters.

Lastly, we have the layered architecture. This one offers the best of both the MVC and hexagonal architectures: we have clear divisions and good abstraction for our core, implementations, and view. Adding more layers to your architecture is simple, and the rules are easy to understand, making it a good solution for teams with developers of different experience levels. Because of good separation and abstractions, you can simply change implementations, and unit testing remains easy.

You now know what the MVC, hexagonal, and layered architectures are, and you learned about the advantages and disadvantages of each implementation. In the next section, you'll briefly learn what design patterns are and how they differ from architectural patterns.

Design patterns used within Angular applications

While architectural patterns focus on a high-level overview of how we segment and abstract our code, design patterns focus on how we implement things within our code. When developing Angular applications, we are already working with some design patterns out of the box. This is because Angular is a strongly opinionated framework with strong **object-orientated programming** (**OOP**) principles at its core. Some design patterns we use by default within Angular applications are the observable, dependency injection, decorator, component, and singleton patterns. These and other design patterns, such as the factory and inheritance patterns, are embedded in how the Angular framework works and should be used throughout your application when you use the framework correctly. Because these patterns are somewhat concealed within the tools and ways of working of the Angular framework, you might be using them without actually understanding how they work at their core. Besides these design patterns embedded within the Angular framework, you can improve your code by introducing even more design patterns. Some work very well in combination with Angular, such as the facade pattern.

As with architectural patterns, design patterns ensure that you adhere to specific rules when implementing your code, resulting in code that is easy to adjust and extend. They prevent you from making the wrong dependencies throughout your code and provide a structured and battle-tested way to approach common problems with software engineering. For now, I wanted to briefly explain design patterns and list some design patterns the Angular framework uses by default. In *Chapter 5*, you will learn about design patterns in more detail. You will learn about different patterns, when to use them, and how to implement them correctly.

You now know what design patterns are and how they differ from architectural patterns. You've learned about some patterns that are used within Angular by design and that you can add more patterns to improve your code implementations throughout your code base. The next topic we will discuss is Nx, which makes structuring, creating, maintaining, and testing large Angular monorepos easy.

What is Nx and why should you use it?

In this section, you will learn what Nx is and why it's such a fantastic tool for developing Angular applications at scale. **Nx** is rapidly becoming the go-to tool for developing large monorepo front-end applications. So, what exactly is Nx and why should you use it?

Nx is a tool that helps you to speed up, streamline, and standardize your development, testing, build, and deployment process. The Nx tooling offers various features and integrations you can utilize during every stage of development. Nx was created so that you can adopt it incrementally by picking and choosing what you want to use or add to your current environment. At its core, Nx helps you with the following:

- Speeding up the build and test times of your applications.
- Managing dependencies and running tasks within monorepo projects.

- Swiftly scaffolding new code snippets, applications, and libraries without needing to worry about configuring build tools.

- Integrating new tools into the projects of your monorepo workspace.

- Ensuring uniformity and consistency within the code of different projects.

- Updating applications and tools through automated code migration.

In the preceding tasks, multiple tools, features, and options, such as a CLI, generators, and plugins, help you achieve your goals and streamline processes. The tools and features Nx has to offer are divided into different modules in their ecosystem.

To start, you have the Nx **command-line interface (CLI)**. Similar to the Angular CLI, it lets you run commands for tasks such as creating workspaces, scaffolding projects, testing projects, or serving and building. Next, the **Nx package** contains all the fundamental technologies Nx offers: task running, workspace analysis, build caching, scaffolding, and automated code migrations. Then, there are plugins, which are NPM packages that extend the fundamentals of Nx and can be created by the Nx community for various purposes, such as generating projects, integrating tools, and adding or updating libraries. Another element of Nx is its **Devkit**, which can be used to build plugins to extend the Nx tooling to your specific needs. Nx also has something called **Nx cloud**, which speeds up your CI with remote caching and distributed task executions, but this is outside the scope of this book. Lastly, we have the **Nx console**, an extension for VS Code, IntelliJ, and VIM, making it much easier to manage your Nx workspace and run Nx commands.

Now that you know what Nx does at its core, let's examine it in more detail and see how it can help you build scalable monorepos for your Angular applications.

How Nx helps you build Angular monorepos that scale

Now that you understand what Nx is at its core, let's dive deeper and explore how it helps you build, test, and standardize your Angular applications. We will begin with one of the main features Nx offers: speeding up tasks such as building, serving, and testing.

Improving build times with computational caching and incremental builds

Typically, when we run tasks with the Angular CLI, such as `ng build` or `ng serve`, our entire application and all the libraries it depends on need to be compiled to complete the build or serve your application. This can become time-consuming as the application grows. The result is slow CI builds and developers waiting for the application to compile each time they want to start or test it. Nx helps to resolve these issues with incremental builds and computational caching.

With computational caching, Nx will check if anything has changed since the last time you ran a command. If nothing has changed or the build computation is equal to a previous cached run, Nx won't rerun the command and instead take the results from its caching system. First, it will look at the local caching, and if you set up remote caching with Nx cloud, it will also check if it can find the same computational hash in the remote cache. If Nx cannot find the same computational hash, it will run the command and store the hash of the result in the Nx cache.

Besides computational caching, Nx helps speed up our build and compile times with incremental builds. When using incremental builds, we only build projects that have been changed since your last build. In a regular scenario, we build the application and all the libraries the application uses. As the application grows and depends on many libraries, it can become time-consuming and costly to rebuild everything each time you build the application. To use incremental builds, your libraries must be buildable so that Nx can cache the libraries and only build them if they changed since your last build. When you're building smaller applications, you might not want buildable libraries with computational caching because making a library buildable has some overhead as well. When you create a library in your Nx workspace, you can choose if you want it to be standard, buildable, or publishable. We will dive deeper into this topic in the *Structuring Angular applications and libraries* section.

Running tasks effectively in a monorepo with Nx

Running tasks, such as `ng build`, `ng test`, and `ng lint`, with the Angular CLI in a single Angular project is straightforward. But things become more complicated when you have a monorepo with tens, hundreds, or even thousands of applications and libraries. In many scenarios, you want to run tasks simultaneously for multiple (or all) projects. Sometimes, you want to run specific tasks when something changes in a project, or you need to know if changes in a library have affected other projects and break them. If you perform these tasks by running commands one by one, this becomes unmanageable quickly, and building tools to watch and check for the affected projects becomes complicated. Luckily, Nx has everything we need to run tasks for multiple or affected projects, watch for changes, and react to them with commands.

When you use the Nx CLI, it is advised not to use the Angular CLI within the monorepo. We want Nx to do all its magic, and this is not possible if we start to generate things with the Angular CLI. Luckily, Nx has you covered!

Let's start with the basics – running commands for a single project using the Nx CLI. Running tasks for a single project is similar to running tasks with the Angular CLI. For example, if we want to run the tests for an application named **testApp**, we can run the following command:

```
nx test testApp
```

If we want to run tasks for multiple projects, we can use the `run-many` keyword combined with the `-p` flag to define the projects for which we want to run the tasks. If you omit the `-p` flag and only use the `run-many` keyword, the task will run for all projects. We can also add the `-t` flag to run multiple tasks at once. For example, if we want to build, lint, and test all projects, we can run the following command:

```
nx run-many -t build lint test
```

Now, let's say we only want to build, lint, and test *testApp* and *testApp2*. For this, we can run the following command:

```
run-many -t build lint test -p testApp testApp2
```

As you can see, running commands with the Nx CLI is simple, even if you need to do it for multiple or all projects within your monorepo. Even if you want to run several tasks simultaneously for a subset or all projects, you can do it quickly and with a single command.

Another helpful option is to watch for changes in specific projects and run scripts whenever changes occur in a watched project. For example, you can watch for changes in an application and echo the project name and the changed filename. This can be done for single projects, a subset of projects, or all projects, like running regular commands. Here's an example of what this command will look like:

```
nx watch --projects=testApp,testApp2 --includeDependentProjects --
echo \$NX_PROJECT_NAME \$NX_FILE_CHANGES
```

Lastly, we need something to detect the affected projects and run commands for them. When a monorepo grows, it becomes too time-consuming to run tests for all projects every time you change a library. Consider the following scenario. We have five applications and three libraries being used by those applications. Now, if we change the code of a library and all five projects use this library, our changes can affect and break the four other projects we aren't working on ourselves. This can be a massive problem if we release an application or library to production without noticing we broke another application with our changes. When we make changes in a library, we need to know what projects in the monorepo are affected by those changes and run the appropriate tests to see if everything still works. For this scenario, Nx has *affected commands*. By running these commands, you can run tasks such as linting and testing for all projects affected by your changes. The following is an example of such a command:

```
nx affected -t test
```

With that, you have basic knowledge of running tasks in an Nx monorepo. Next, let's explore how Nx helps us keep the code and setup in our monorepo projects standardized and uniform.

Ensuring uniformity and consistency within your Nx monorepo

Another critical aspect of a scalable Angular monorepo is uniformity and consistency in the code of the different applications within the monorepo. When you have hundreds of projects in your monorepo, you don't want different code conventions and implementations in each project. This would make it much harder for developers to start working on other projects within the monorepo, and it also makes it harder to find similar code or refactor code in bulk. When everyone uses the same conventions and code patterns, each developer can work on every project within the monorepo and make changes when needed.

Nx helps us maintain uniformity with global linting rules that apply to all projects within the monorepo. You can set up these linting rules however you see fit and can even create custom linting rules. Besides global linting rules, you can also apply specific lint rules for individual projects.

Nx generators are another great way to enforce uniformity within your Nx monorepo. Generators are used to generate applications, libraries, components, and other code snippets. You can even use Nx generators to modify your code throughout the monorepo or set up and change configuration files. This can be very useful when you need to apply refactoring to multiple files of your monorepo.

Using generators ensures that things are created in the same manner by all developers working within the monorepo. Executing generators can be done with the terminal, though when using the Nx console, you get a nice user interface for executing them. You can also overwrite the built-in generators so that fewer options are revealed, and the default values are set. This reduces variation and makes the generators easier to use for less experienced developers. We will dive deeper into this topic when we create an Nx monorepo in the *Setting up a scalable Angular workspace* section. For now, it's important to know that you must use generators whenever possible.

The last thing I want to mention on uniformity is **Prettier**. Nx has built-in support for the code formatter Prettier. As with linting rules, formatting rules can be configured for the entire monorepo and individual projects within the monorepo. You can also run a command with the Nx CLI to detect unformatted lines of code. When you configure VS Code to auto-format on **Save**, Prettier will format your code whenever you save a file.

Now that you've learned how to manage monorepos with Nx, we will explore how you can update and manage dependencies in an Nx monorepo.

Nx dependency management and code migrations

Managing dependencies in an Nx monorepo is easy since there is one `package.json` file for the entire monorepo (this is only the case for integrated monorepos, but we'll cover this in more detail when we create our monorepo in the *Setting up a scalable Angular workspace* section) and most packages can be updated automatically, including configuration and code changes needed for that update. Running automated updates can be done by using the `nx migrate` command, like so:

```
nx migrate [packageAndVersion]
```

In general, you will update to the latest version of the Nx package and dependencies; in that case, you can use the following command:

```
nx migrate latest
```

Running the preceding command will update your package.json file and generate a migrations.json file containing all the migrations that Nx and its plugins need to run to update all packages successfully. To execute these migrations, you need to run the migrate command with an additional flag, like so:

```
nx migrate --run-migrations
```

After running this command, your packages, configuration files, and code will be updated to use the latest versions of Nx and the plugins configured within the workspace. For packages that don't have a plugin, you need to do the updates manually as you do with NPM packages. Packages with a plugin will do everything for you with Nx migrations. There are plugins for Angular, Jest, Cypress, and most other things you will use within the Nx monorepo.

Another cool feature for managing dependencies is that Nx lets you visualize dependencies on a graph. With this, you can see what projects depend on one another and zoom in on these nodes to better view the dependencies. This works for projects as well as tasks. If you have tasks that run multiple tasks in a chain, you can also visualize this task chain on a nice graph. The **Nx graph** is a valuable tool as visualization helps you better understand your monorepo's structure. If you can no longer make sense of the dependencies in the graph, chances are that they won't make sense in your code either. You can open the graph by running the following command:

```
nx graph
```

If you want a graph of the affected projects, you can run the following command:

```
nx affected:graph
```

Lastly, you can use tags to set up boundaries throughout your monorepo so that projects can only import what they're supposed to import. If you try to import a library you aren't supposed to, the code won't compile, and an error will inform you that you aren't allowed to use the library within your project. When we create our monorepo in the *Setting up a scalable Angular workspace* section, we'll set up these boundaries as well.

As mentioned previously, Nx is a vast topic that offers us many tools for managing monorepos. As we progress through this book, we will learn more about Nx, but for now, you have a good idea of what Nx is and how it can help you manage Angular monorepos of any size. Now, we will move on to creating an Nx monorepo, including some Angular projects and libraries.

Setting up a scalable Angular workspace

In this final section, we will create an Nx monorepo, along with some placeholder Angular applications and libraries for demonstration purposes. We'll start by adding some extensions in VS Code to make our developer experience more pleasant.

Add the following extensions:

- Angular Essentials (version 16): Angular Essentials is comprised of eight different extensions that are useful for Angular development

- Angular support

- Nx console

Once you've installed these extensions, you will probably need to reload VS Code for all of them to be applied, so restart your VS Code application. After this, you should see an Nx icon on the left-hand side of VS Code; this is the Nx console. When you have successfully installed the various VS Code extensions, you must globally install the Nx and Cypress NPM packages. To do so, you can run the following command in the terminal of your choice. I like to use the integrated VS Code terminal:

```
npm i -g nx cypress
```

Once these NPM packages have been installed, we can create our Nx workspace. Find a folder where you want to create the Nx monorepo and open a terminal at this location. To create the Nx workspace, run the following command:

```
npx create-nx-workspace
```

When you run this command, you will be prompted with several questions:

1. Where would you like to create your workspace? (`business-tools-monorepo`.)

2. Which stack do you want to use? (Angular: Configures an Angular app with modern tooling.)

3. Standalone project or integrated monorepo? (Integrated monorepo: Nx creates a monorepo that contains multiple projects.)

4. Application name (`Invoicing`).

5. Which bundler would you like to use? (`esbuild`.)

6. Default stylesheet format (SASS (`.scss`)).

7. Do you want to enable **server-side rendering (SSR)** and **static site generation (SSG)**? (No.)

8. Test runner to use for **end-to-end (e2e)** tests (Cypress).

9. Set up CI with caching, distribution, and test deflaking (Skip).

10. Would you like remote caching to make your build faster? (Yes.)

Once you've answered these questions, the Nx monorepo will be created in a folder called `business-tools-monorepo`. We will dive into what Nx created for us in the *What did Nx create in our new workspace?* section, but first, I want to explain the different types of workspaces you can create with Nx.

Nx workspaces come as standalone projects, package-based monorepos, and integrated monorepos:

- Standalone projects are well suited for when you want to start with a single project but keep the door open for growing into a large monorepo. Nx is also useful for single projects as you can utilize their generators and automated updates through Nx migrations and executors.

- Next, we have package-based monorepos. This solution is beneficial when you need to add a lot of exciting projects to an Nx workspace. With package-based repos, each project in the Nx workspace has its own dependencies. You will get the improved speed and task running from Nx, but Nx will stay out of your way for the rest.

- Lastly, there are integrated Nx monorepos. This is what we'll be using and is the setup where you leverage all the tools Nx has to offer.

Now, let's see what Nx created f when it generated the monorepo.

What did Nx create in our new workspace?

When you open the `business-tools-monorepo` folder in V SCode, you will see five folders and a couple of files in the root of the monorepo. We will review the essential files and explain the folders, but first, let's briefly lay out what Nx did for us when it generated the monorepo.

Nx created our monorepo and configured it for building Angular applications. It installed the Nx Angular plugin and all the essential NPM packages needed to develop Angular applications. Besides that, Nx configured everything for computational caching, ESlint, Cypress e2e testing, and unit testing with Jest, and Prettier for code formatting. Nx also made our first Angular application, **invoicing**, in the `apps` folder. The `apps` folder will contain all the applications we create in our Nx monorepo. If you open the `apps` folder, you will notice that Nx also created an `invoicing-e2e` project. This is a Cypress project for e2e testing our Angular invoicing application. All of these things will work out of the box. Nx did all the configuration for us. In the folder structure, we'll see a folder named `apps`.

Now, let's look at some important files that Nx created in our workspace:

- `nx.json`: This is used to configure the Nx CLI and its default configurations
- `.eslintrc.json`: This file can be found at the root of your monorepo and inside each project in the repo. The file in the root contains the global ESlint rules that are applied to all projects within our monorepo. The files within the specific projects contain project-specific ESlint rules.

- `project.json`: This can be found at the root of our Angular application (`apps\invoicing\project.json`). In a regular Angular application, this would be our `angular.json` file. It has the same content and usage. The `project.json` file will be created for each project in our Nx monorepo.

- `tsconfig.base.json`: This is used for global TypeScript settings and setting up aliases for library imports.

We will learn more about these files and how to edit them throughout this book and when we start to add projects, libraries, and plugins to our Nx monorepo.

Improving our Nx workspace for better Angular development

We already have a nice workspace from Nx that will scale pretty well out of the box. You can uniformly create Angular applications, libraries, and other classes throughout the workspace. For each application or library you create, Jest unit testing is configured, and for each application, you get a Cypress project with everything you need to run e2e tests. ESlint and Prettier are configured for the entire monorepo to keep your code uniform among the different projects. The monorepo has automated updates with Nx migrations; you can visualize your dependencies and tasks with the Nx graph and effectively run commands against your monorepo with the Nx task runner. The first step I like to take is creating a `libs` folder next to the `apps` folder. Inside this `libs` folder, we will store all our libraries. To improve this further, we need to add some extra ESlint rules, create custom generators tailored to our needs, and set up project boundaries with tags.

Adding ESlint rules to improve our Angular code

We can use lint rules to keep our code uniform and ensure that developers don't implement code in unintended ways. Linting can catch bugs preemptively and enforce best practices. Nx has already added Angular-specific lint rules for our monorepo with two `@angular-eslint` packages. To make our monorepo more robust, we'll want to add some extra ESlint rules tailored for Angular and RxJS. The lint rules we will add are not required, but I suggest using them or adding rules according to your preferences when setting up a monorepo. This is important so that everyone working on your code will follow the same implementation and style guide rules for writing code. Each company and code base uses its own conventions, so make sure you have lint rules in place to enforce them. If you can't find lint rules that enforce your code conventions, Nx also allows you to add custom lint rules.

First, we will add some NPM packages as dev dependencies:

```
npm i --save-dev eslint-plugin-deprecation eslint-plugin-rxjs eslint-plugin-rxjs-angular
```

When the command has finished running, the packages will be added to the `package.json` file in the root of your Nx monorepo, and the lint rules belonging to these packages can be added to the `.eslintrc.json` files. We will only add rules to the `.eslintrc.json` file at the root of the monorepo. These lint rules will be applied to all the projects within the monorepo. We also need to add `parserOptions` to the `.eslintrc.json` file because we're working with TypeScript and type-based lint rules. You can find the updated `.eslintrc.json` file in this book's GitHub repository (link in the *Technical requirements* section).

Once you've added the extra lint rules and `parserOptions`, we can move on and start creating some custom Nx generators.

Creating custom Nx generators for our Angular monorepo

Generators are excellent tools to do things uniformly because everyone who uses the generator will have the same result. First, we'll focus on overwriting built-in Angular generators. Throughout this book, we will create and use generators to refactor code in our monorepo and to create custom code snippets.

In this section, we will overwrite the generator for creating Angular libraries. We will significantly reduce the options you can enter when using the generator, add custom options to enforce a directory structure, and add tags for project boundaries. Fewer options means less deviation from the conventions your organization is using, resulting in more uniformity and fewer questions from less experienced developers.

Let's start by creating an Nx plugin using an Nx generator:

1. To use the plugin generator, install the NPM package by running the following command:

    ```
    npm i @nx/plugin
    ```

2. Now, on the left-hand side of VS Code, click on the Nx icon to open the Nx console.

3. Go to the **GENERATE & RUN TARGET** tab and click **generate**. This will open a dropdown with a search bar at the top of VS Code.

4. In this search bar, enter `plugin` and select the **@nx/plugin - plugin Create a Nx plugin** option. This will open a new window in VS Code where you can generate your plugin.

5. By default, the generator will ask you to fill out two fields, a *name* and an *import path*, and a directory, of which the name is required.

6. Name your plugin `workspace-generators-plugin`. For the import path, enter `@business-tools/workspace-generators-plugin`, and for the directory, enter `libs`.

7. Click **Show all options**.

8. Under **projectNameAndRootFormat**, select **derived**.

9. Then, click on the **generate** button at the top right.

This will generate your plugin under the `libs` folder at the root of your monorepo.

Next, we will create our custom generator:

1. First, click on **generate** again in the Nx console.

2. In the search bar, type `generator` and select **@nx/plugin** - **generator**.

3. Now, in the newly opened window, give the generator a name, such as `generate-angular-library`.

4. For **directory**, enter `libs/workspace-generators-plugin`.

5. Click **Show all options**.

6. Under **projectNameAndRootFormat**, select **derived**.

7. Then, click on the **generate** button at the top right.

When this process completes, you will find a `generators` folder inside the `src` folder of `workspace-generators-plugin`. Inside this `generators` folder, you will find your custom generator, which is named `generate-angular-library`. Inside your custom generator are a bunch of files, but before we start exploring them, let's look at what we will overwrite with this custom generator:

1. Go back to your Nx console and click **generate** again.

2. This time, search for `library` and choose **option @nx/angular** - **library**.

3. When you inspect the new generator window for this library, you will find seven options to fill out; if you click on **Show all options**, you will have 30 possibilities to fill out. Not all developers will know what to select here, and if we leave it up to developers, we will get too much variation when we create libraries within our monorepo. For now, let's close the window for generating an Angular library and start to overwrite it with our custom generator.

When we overwrite an existing generator, we are interested in three different files: `generator.ts`, `schema.json`, and `schema.d.ts`. We will write our logic inside the `generator.ts` file; the `schema.json` file will contain the information for the Nx console window, and the `schema.d.ts` will contain the interface for our options inside the `generator.ts` file. Let's inspect these three files individually, starting with the `generator.ts` file.

Nx has already generated a bunch of code for us in this `generator.ts` file, but we won't worry about all that and instead replace it with the following small piece of code:

```
import { Tree } from '@nx/devkit';
import { libraryGenerator } from '@nx/angular/generators';
import { GenerateAngularLibraryGeneratorSchema } from './schema';

export async function generateAngularLibraryGenerator(
  tree: Tree,
  options: GenerateAngularLibraryGeneratorSchema
) {
```

```
    await libraryGenerator(tree, options);
 }
 export default generateAngularLibraryGenerator;
```

We cleared everything inside the generateAngularLibraryGenerator function and replaced it with a libraryGenerator method from @nx/angular/generators. This libraryGenerator method is the built-in Angular library generator we inspected a moment ago. When you want to overwrite built-in generators, you can find them in your node_modules folder or the GitHub repository of the corresponding package. Now, if we build and run our custom generator, we can only enter a name because that is the only thing we have in our schema.json file, so our custom generator generates an Angular library for us with the default settings. So, we went from 30 options to fill out to only a name. However, we want to add additional options and choose specific values for some of the built-in options. We can do this by editing the libraryGenerator method, like so:

```
 await libraryGenerator(tree, {
   name: options.name,
   simpleName: true,
   standalone: true,
   buildable: true,
   prefix: `bt-libs-${options.type}`,
   style: 'scss',
   changeDetection: 'OnPush',
   directory: `libs/${options.domain}/${options.type}`,
   tags: `domain:${options.domain}, type:${options.type}`,
   importPath: `@bt-libs/${options.domain}/${options.type}/${options.
 name}`,
 });
```

Our generated libraries will now be buildable, use standalone components, have on-push change detection, use SCSS style, have tags, have an import path, and have a directory structure configured. We use a name, domain, and type property in our options object. The name is configured by default, so let's add the options we want to expose inside our schema.json and schema.d.ts files.

Inside the schema.d.ts file, we'll find an interface for the options object we used in the generator.ts file. This interface declares a name property by default. We want to add a domain and type to the interface and only allow preset string values for these properties. We can do this by editing the interface like this:

```
 export interface GenerateAngularLibraryGeneratorSchema {
   name: string;
   domain: 'finance' | 'hr' | 'marketing' | 'inventory' | 'shared' ;
   type: 'ui' | 'data-access' | 'feature' | 'util' | 'all' ;
 }
```

Lastly, we want to update our `schema.json` file. This `schema.json` file defines the values we can fill out in the Nx console or terminal when using the generator. When we generated our custom generator, Nx added a name property for us, so we must add the domain and type. We can do this by adding extra objects in the `properties` object inside the `schema.json` file. To include the domain, you can add the following code:

```
"domain": {
  "type": "string",
  "description": "Domain of the library",
  "$default": {
    "$source": "argv",
    "index": 1
  },
  "x-prompt": {
    "message": "What domain would you like to use?",
    "type": "list",
    "items": ["finance", "hr", "marketing", "inventory", "shared"]
  }
}
```

For the `type` field, you can add a similar object under the `domain` object. The `index` property needs to be one unit higher for the `type` object because this index indicates at what position the field will be shown in the Nx console when we use the generator.

Now, if we build and run the generator, we'll have three fields to fill out. When we fill those fields, a library will be generated for us with all the configurations we added.

But we can improve the generator even more. We can do this by adding an `'all'` option to the type. When selecting `'all'`, we generate a library for each type at once, and we can add cleanup logic to remove the initial component that comes with the generator – in most cases, we don't want a component named after the library. To achieve these improvements, we'll need to extract the `libraryGenerator` method into a separate function underneath the `generateAngularLibraryGenerator` function:

```
async function generateLibrary(
  tree: Tree,
  options: GenerateAngularLibraryGeneratorSchema,
  type: string
) {
  await libraryGenerator(tree, {
    name: options.name,
    simpleName: true,
    standalone: true,
    buildable: true,
```

```
    prefix: `bt-libs-${type}`,
    style: 'scss',
    changeDetection: 'OnPush',
    directory: `${options.domain}/${type}`,
    tags: `domain:${options.domain}, type:${type}`,
    importPath: `@bt-libs/${options.domain}/${type}/${options.name}`,
  });
}
```

Next, we need to add an array with our types above the `generateAngularLibraryGenerator` function, like this:

```
const TYPES = ['ui', 'data-access', 'feature', 'util'];

export async function generateAngularLibraryGenerator(
  tree: Tree,
  options: GenerateAngularLibraryGeneratorSchema
) {
  ..........
}
```

Now, inside the `generateAngularLibraryGenerator` function, we'll replace `libraryGenerator` with a `for` loop. This will add a library for each of the available types when `options.type` is equal to `'all'`:

```
    if (options.type === <all>) {
      for (const type of TYPES) {
        await generateLibrary(tree, options, type);
      }
    } else {
      await generateLibrary(tree, options, options.type);
    }
```

The only thing that is left to do now is to add some logic to clean up the initial component files and update the `index.ts` file to remove the export of this initial component. Otherwise, if we create a library named `common-components`, for example, it will create a component named `common-components-component`, and that is not something we want.

To clean up our library, we will use the `tree` parameter that's exposed to us by the `generateAngularLibraryGenerator` function. This `tree` object contains our monorepo tree and, in it, all our folders and files. To remove the initial component files and update the `index.ts` file when we generate a library, we can add the following code at the bottom of our `generateAngularLibraryGenerator` function:

```
const path = `libs/${options.domain}/${options.type}/${options.
name}/src`;

tree.delete(`${path}/lib/${options.name}`);
tree.write(`${path}/index.ts`, <>);
```

Once you've updated and saved everything, you can test the generator.

1. Open the Nx console in VS Code. At the top, you will find a section named `projects`.

2. Underneath this section, you will find your **workspace-generators-plugin** project. If you expand this, you will see three options: **build**, **lint**, and **test**.

3. When you hover over the **build** option, you'll see a **play** button. Go ahead and click on this button to build the **workspace-generators-plugin** project.

4. When the plugin is built successfully, restart VS Code and click on **generate** in your Nx console. You should see **@business-tools/workspace-generators-plugin - generate-angular-library generator**.

You only need to restart VS Code when you add a new generator or adjust `schema.json` because Nx schematics are loaded when you open VS Code.

When we use the generator, there are three fields where we can enter the name, domain, and type. Let's create a library called `common-components` and then select **shared** for the domain and **UI** for the type. This will generate a library for us at the `libs\shared\ui\common-components` path.

We will also overwrite the component generator. You can go ahead and try to do this yourself. If it still feels challenging, you can go to this book's GitHub repository and take the code from there (link in the *Technical requirements* section). If you did the overwrite on your own, please check if you implemented it the same as we did so that you can continue using the same generators throughout this book.

With that, you know how to overwrite generators and have created your first custom generator to build Angular libraries. Next, we will create project boundaries for the Nx monorepo.

Setting up project boundaries for the Angular monorepo

When you work in a large monorepo where different teams work on separate projects, it is essential to have some boundaries. If you have a library only intended for one domain of the monorepo, you don't want someone to create unintended dependencies by importing that library into another domain.

For example, if you have a library intended for finance-related applications, you don't want it to be imported into a marketing application and create a dependency between our finance and marketing domains. What's even worse is if someone made a direct dependency between the two applications. If we do this, we have to build and deploy both applications when we only want to update and deploy one of them.

With Nx, we can set up boundaries with **tags**. The tags that belong to a project can be found inside its `project.json` file. The boundaries are defined inside the `.eslintrc.json` files. We will only create global boundaries in the root `.eslintrc.json` file. Inside this `.eslintrc.json` file, you will find the following lint rule:

```
"@nx/enforce-module-boundaries"
  "error",
  {
    "enforceBuildableLibDependency": true,
    "depConstraints": [
      {
        "sourceTag": "*",
        "onlyDependOnLibsWithTags": ["*"]
      }
    ]
  }
```

You can update the `depConstraints` array to update the boundaries for your Nx monorepo. By default, the object inside this array uses * as a wildcard, allowing any project to import every other project. I advise removing this object and setting strict project boundaries. For example, if you only want projects with domain finance to import other projects with the same domain, you can add this object to the `depConstraints` array:

```
{
  "sourceTag": "domain:finance",
  "onlyDependOnLibsWithTags": ["domain:finance"]
}
```

The `sourceTag` component defines the tag we are targeting, while `onlyDependOnLibsWithTags` defines the tag it is allowed to import. If you want to see the constraints I configured for the monorepo, you can take them from this book's GitHub repository, but you are free to set up the constraints however you like.

After you've configured the boundaries, you will get a lint error in VS Code if you try to import something you're not allowed to. After the initial setup, you might need to restart VS Code for the boundaries to take effect without running the linter.

Now that we've improved our Nx monorepo and it is ready to host many Angular applications and libraries, let's wrap things up by discussing a good structure for our Angular applications and libraries within our monorepo.

Structuring Angular applications and libraries

The last thing we must cover is how to structure our Angular applications and libraries within the Nx monorepo. Seeing your applications as containers that build up their pages with the components and logic from libraries is a good practice. This incentivizes a modular approach and makes it easier for you to have a good separation of concerns as a dedicated library project is a much greater boundary than separating code with folders in your application. Generally speaking, you can use the 80/20 rule, where 80% of your code lives within dedicated library projects and 20% within your application projects.

These libraries don't have to be built separately from the projects that consume them. If that is the case, everything stays the same regarding your deployment process, but you won't utilize Nx incremental builds and computational caching. If you want to use incremental builds or publish your libraries to an external registry such as NPM, you can mark them as buildable or publishable when you generate them. In our custom generator, we made the libraries buildable by default. For small applications, you can consider using regular libraries that aren't buildable unless you want to publish them.

Placing code in a library doesn't necessarily mean the code has to be general purpose and must be consumed by multiple parties; putting code in an Nx library can be purely for organizational goals. It stimulates you to think in a more API-driven way about your code, often resulting in cleaner implementations with fewer dependencies, which, in turn, might result in code that can be reused, but it doesn't have to be. When organizing libraries and applications, you should consider the different business domains. Generally speaking, the teams of an organization are aligned with these business domains; thus, it makes sense to have a similar organization for your monorepo projects.

Do you create a new library or reuse an existing one?

The decision of when to create a new application is pretty straightforward. It should probably be an application if it's a product with a UI that can be used or sold independently. For libraries, it can sometimes be harder to define when to create a new library and when to add code to an existing library. As with most programming decisions, whether to start or reuse a library is about trade-offs.

The main benefits of creating a new library and splitting up your code are that tasks such as building and testing will be completed faster, you have better visualization of your architecture with the Nx graph, and you have more control over project boundaries by using tags. The advantages of reusing a library are that you can better group related code together without any constraints, so it's easier to experiment and less prone to mistakes. Especially when your code base is rapidly evolving, keeping things in one library might be easier for the time being. A good practice is splitting things into multiple libraries when the development pace slows.

Libraries are generally divided into UI, feature, utility, and data-access libraries. Your UI libraries should only contain dumb presentational components. Feature libraries contain smart components with access to your facade services and are created for business cases or pages within the applications. The data-access libraries house state management logic and provide everything to communicate with the back-end APIs. Lastly, we have the utility libraries, which host helper functions and other useful low-level utilities.

All your libraries should be grouped by their respective application or a business domain containing multiple applications. For large companies, the libraries can be grouped by sections of a specific business domain, depending on how the scope of each domain is defined. When you want to move an application or library into a new folder within your Nx monorepo, you need to use the Nx move generator by running the following command:

```
nx g move --project some-library target/folder/path
```

If you want to remove an application or library, you should use the Nx remove generator, like so:

```
nx g remove some-project-name
```

For both moving and removing projects, you can use the Nx console for a more visual approach. It is important to use the Nx console or run the respective Nx command in the terminal to perform these actions because they will automatically update all configuration files within your Nx monorepo.

Figure 1.5 shows a proposed folder structure, although this is entirely up to you and the structure and needs of what you're building:

```
apps/
  └── finance/                          <---- business domain
        └── invoicing/                  <---- application
libs/
  ├── finance/                          <---- business domain
  │   ├── ui/                           <---- library type
  │   │   ├── invoice-components        <---- library
  │   │   └── accounting-components
  │   ├── feature/
  │   ├── data-access/
  │   └── util/
  └── marketing/                        <---- business domain
        ├── ui/                         <---- library type
        ├── feature/
        ├── data-access/
        └── util/
```

Figure 1.5: Folder structure

Before moving on to the next chapter, let's clean up our monorepo and add some placeholder projects. First, remove the invoicing application we made earlier and its corresponding e2e project. You can do this by right-clicking on the project and selecting **Remove Nx project**. This will open an Nx console window for removing projects. In the dropdown, you can choose the project you want to remove. First, remove the `invoicing-e2e` project; then, remove the `invoicing` project. Once you've removed both projects, we can create some placeholder applications with the custom application generator we took from this book's GitHub repository.

Create the following projects:

- `expenses-registration` (under the domain finance)
- `social-media-dashboard` (under the domain marketing)

The `placeholder` library we created can stay because it is already equipped with tags and a domain and was created with a custom generator. We will create extra libraries as we continue and start to add some code.

Summary

In this chapter, you created an Nx monorepo with some placeholder projects and are now ready to scale to hundreds of applications. You learned how to structure your Angular applications and libraries within the monorepo, and now know how to utilize the essential features of Nx.

In the next chapter, we will explore some of the newest and most powerful features in the Angular framework.

2
Powerful Angular Features

Angular has built-in tools for everything you need to build robust web applications. In this chapter, you will learn about the newest and most powerful features in the Angular framework. We will also dive into Angular component communication, the router, and arguably the most important and powerful part of the Angular framework: dependency injection. By the end of this chapter, you will know how to inject, consume, provide, and adjust the hierarchy of dependencies, how to communicate between components, and how to effectively use the Angular router.

This chapter will cover the following main topics:

- What makes Angular so powerful?
- New features in the Angular framework
- A deep dive into the Angular router
- Component communication
- Dependency injection

What makes Angular so powerful?

Angular sets itself apart from other popular frontend frameworks and libraries such as ReactJS and VueJS because it is strongly opinionated and has everything you need to develop complex web applications embedded in the framework itself.

Using Angular effectively ensures that there is a certain level of consistency and that best practices are implemented. This is because Angular makes many decisions for you, such as using TypeScript and relying on **object-oriented programming** (**OOP**) principles and built-in tools to handle common problems such as routing, HTTP requests, and testing! Because Angular has everything built into it, you don't need to bring in a lot of external packages, reducing the surface for potential exploits or packages that stop being maintained. These aspects often make Angular the framework of choice when building complex frontend systems or enterprise software composed of multiple applications.

Angular comes packed with powerful and useful features for building web applications. If you are reading this book, you should already be familiar with the main features, but we will still mention the ones that are most essential to the framework:

- **Components and services**: Components and services are the building blocks of Angular applications. Components are used to develop reusable UI elements and pages comprised of these UI elements. Angular services are injected throughout your applications using dependency injection and communicate with backend APIs, handle state management, provide data, and implement business logic.

- **Dependency injection**: Angular dependency injection is one of the fundamental concepts of the framework and is often regarded as its most powerful feature. Dependency injection allows you to inject values and logic throughout your applications.

- **Signals**: Signals are a new concept within the Angular framework that is used to define stateful properties. Angular tracks where and how the Signal values are used to optimize change detection, resulting in improved performance. Signals are also reactive, allowing you to automatically react when a Signal value changes.

- **HTTP client**: The built-in HTTP client provides an elegant and intuitive interface for communicating with APIs and fetching data. With built-in features such as request and response interceptors, error handling, and observable-based responses, the Angular HTTP client is everything you need for handling HTTP logic.

- **Data binding**: Angular data binding enables real-time synchronization between the `component` class and its corresponding template, facilitating data updates without manual intervention. In the Angular framework, data binding can be done in three ways: from the component class to the view by using the square bracket notation, from the view to the component class with events and the round bracket notation, and two-way data binding with the square and round bracket notation, also known as banana-in-a-box.

- **Router**: The Angular Router facilitates the creation of single-page applications with dynamic routing capabilities. It enables developers to define routes and associate them with specific components, allowing seamless navigation between different views and pages within the application.

- **Directives**: Angular directives are an essential building block of the framework and allow you to extend HTML elements with additional functionality and behavior or add and remove DOM elements.

- **Pipes**: Pipes are used for transforming and formatting data within HTML templates. Using pipes helps to maintain clean and concise templates while avoiding excessive logic in the component code.

- **Forms**: Forms are at the heart of each web application. Angular forms come in two types: **template-driven forms** and **reactive forms**. With features such as validation, error handling, and data synchronization, forms help you to develop robust applications.

These features are just some of the most powerful features the framework offers. Let's move on and explore what the Angular team changed in the latest versions. The framework is changing rapidly, and new concepts and tools are being introduced to make Angular even more powerful and future-proof.

New features in the Angular framework

The world of web development is evolving rapidly, and because of that, frameworks such as Angular have to keep growing to stay relevant. In this section, we will explore what's new in the Angular framework and why these changes are being made.

Standalone components

In Angular version 14, **standalone components** were introduced as a developer's preview; version 15 released them for production usage. The Angular team introduced standalone components to simplify how we build Angular applications. Before standalone components, everything had to be declared in ngModules. Many developers dislike ngModules, and errors related to ngModules can be hard to resolve.

With standalone components, you can build applications without ngModules. Components, directives, and pipes can be marked as standalone, and then they don't have to be declared in an ngModule.

In *Chapter 1*, we made Nx generators for libraries and applications and applied the standalone flag for both. Because of that, when we generated our Angular applications, they were created with standalone components. Let's look at the decorator from the `app.component.ts` file in your `expenses-registration` application:

```
@Component({
   standalone: true,
   imports: [RouterModule], ………
})
```

If you compare this to a non-standalone component, you might notice two things that are out of the ordinary: the **standalone** flag and the **imports** array. When the `standalone` flag is set to `true`, the component becomes standalone. When something is standalone, it must import all its dependencies directly in the `imports` array instead of getting dependencies from an ngModule. For example, in `app.component.html`, we use the router outlet in the template, and because of that, the component needs to import `RouterModule`; you can also only import `RouterOutlet` instead.

You can also create applications without any ngModules. To do this, bootstrap the application with a standalone component instead of an ngModule. We need to look at our `main.ts` file to see how this is done. In an application that uses ngModules, you will find something like this in the `main.ts` file:

```
platformBrowserDynamic().bootstrapModule(AppModule).catch()
```

In this scenario, the Angular application bootstraps with the AppModule, the root ngModule. When you want to work without ngModules, you can change this and bootstrap with your root standalone component instead. If you look at the `main.ts` file of your *expenses-registration application*, you will find the following:

```
bootstrapApplication(AppComponent, appConfig).catch(......);
```

As you can see, we use the `bootstrapApplication` function instead of `bootstrapModule`. We provide the `bootstrapApplication` function with a standalone component and a configuration object. This configuration object configures things such as routing, the HTTP client, and third-party modules. We will come back to this configuration object a couple of times. For now, remember you can configure application settings here when bootstrapping with a standalone component.

You can also mix ngModules with standalone components. If you have an application that already uses ngModules, you can start using standalone components alongside your existing code and leave your modules as-is.

Now that you understand the basics of standalone components and how to use them, let's explore the new `inject` function for dependency injection.

Dependency injection using the inject function

Another cool feature that was introduced in Angular 14 is the `inject` function, which is used as an alternative for constructor dependency injection. Up until now, constructor dependency injection was the only way to inject dependencies into your Angular applications:

```
constructor(private userService: UserService) {}
```

With the `inject` function, we have an alternative approach that looks like this:

```
private userService = inject(UserService);
```

When we reach the *Dependency injection* section, we will dive deeper into this new syntax. For now, we will move on to the next new feature: directive composition.

Directive composition

Directive composition was a much-requested feature by the Angular community and was added to the framework in version 15. With directive composition, you can apply directives to components by configuring them in the `component` decorator. Each time you use the component in a template, the configured directives will be applied automatically without adding the directive to the HTML element. You can also use directive composition inside other directives, resulting in a directive that applies multiple directives. In *Chapter 3*, we will dive deeper into the topic of directive composition.

For now, you just need to know that you can configure directives inside the component decorators to share common behavior and reduce template complexity. Now that you know about directive composition, let's explore Angular Signals a bit.

Angular Signals

Angular Signals were introduced in Angular 16, and it's one of the most significant changes for the framework since it went from AngularJS to Angular. In Angular 17, the framework also introduced Signal component inputs, and query Signals to query template elements using Signals. With Signals, we have a **reactive primitive** in the Angular framework with which we can manage the application state.

Signals allow you to declare, compute, mutate, and consume values reactively, meaning the Signal will automatically notify all consumers when the Signal's value changes. Because Signals are reactive, you can automatically react when a Signal's value changes, performing logic or updating other values when a Signal is set with a new value. Signals wrap around values and expose them through a getter, which allows the Angular framework to track who is consuming the Signal and notify the consumers when the value changes. Signals can wrap around simple values or complex data structures and can be writable or read-only. Here's a straightforward example of a Signal and a computed Signal value:

```
@Component({ ........ , template: `
    <div>Count: {{count()}}</div>
    <div>Double: {{double()}}</div>`
})
export class AppComponent {
  count = signal(10);
  double = computed(() => this.count() * 2);
}
```

In the preceding example, we have a `count signal` value and a computed Signal that doubles the count. The computed Signal will automatically compute a new value when the `count signal` value changes. To better explain the advantages of Signals, let's define what a reactive primitive is. In JavaScript, you have primitive and non-primitive values. The JavaScript primitives are `string`, `number`, `bigint`, `boolean`, `symbol`, `null`, and `undefined`. The non-primitives are objects.

JavaScript non-primitives are reference types, meaning that if you assign them to a new variable, you don't create a new object but make a reference to the existing object. Primitives don't work like that; if you assign a string to a new variable, it doesn't hold a reference to the original variable but it does create a new string. Primitives are immutable, and non-primitives are mutable. This means you can change a non-primitive after it's created. If you reassign a string, it's a new string and not the same string with a different value. When you adjust an object, it remains the same object, only with different values.

A reactive primitive is an immutable value that alerts consumers when it's set with a new value. All consumers can automatically track and react to changes in this reactive primitive.

> **Important note**
>
> The Signal itself is a reactive primitive and is immutable. You can only update the Signal and notify consumers of the Signal by using the `set()` or `update()` method on it.
>
> Yet, the value held by the Signal is not immutable! So, if you use a non-primitive (an object or array) as a Signal value, you can still mutate the value without updating the Signal itself.

Because Signals are reactive primitives, the Angular framework can better detect changes and optimize change detection and rendering, resulting in better performance. Signals are the first step to an Angular version with fully fine-grained change detection that doesn't need `Zone.js` to detect changes based on browser events. At the time of writing, Angular assumes that any triggered browser event handler can change any data bound to a template. Because of that, each time a browser event is triggered, Angular checks the entire component tree for changes because it can't detect them in a fine-grained manner. This is a significant drain on resources and impacts performance negatively.

In *Chapter 7*, we will dive deeper into Signals and look at different implementations and how you can combine them with RxJS. For now, you just need to know that Signals can be used to manage application state and that they introduce a reactive primitive, which can significantly improve the reactivity and performance of your Angular applications.

Now that you know what Signals are and why they are important, let's learn about the new Angular control flow system.

Angular control flow

The control flow system was introduced in Angular 17 and provides a new mechanism to show, hide, and repeat elements inside HTML templates. Until Angular 17, you could only use the `*ngIf`, `*ngFor`, and `*ngSwitch` directives to show, hide, or repeat elements inside the HTML templates.

As of Angular 17, you can use both the directives and the new control flow system interchangeably. Let's look at an example for each option using the new control flow syntax, starting with the `@if` control flow:

```
@if (a > b) {
  {{a}} is greater than {{b}}
} @else if (b > a) {
  {{a}} is less than {{b}}
} @else {
  {{a}} is equal to {{b}}
}
```

As you can see, the new control flow syntax uses @if, @else if, and @else to define if-else statements inside your HTML templates. The new control flow syntax makes it a lot easier to create if-else statements. You can use both the control flow and the directive syntax, so pick whichever you and your team prefer. Now that you've seen an example of the @if control flow, let's see how you can repeat elements inside the template using the new control flow syntax:

```
@for (item of items; track item.id) {
  <li> {{ item.name }}</li>
} @empty {
  <li> There are no items.</li>
}
```

The @for syntax can be used interchangeably with the *ngFor directive. The new control flow syntax requires you to define a track property. You assign the track property with a unique identifier such as an ID, which allows Angular to only re-render changed items when the list you are rendering changes. You can also provide an @empty block to display something when the array provided to the @for block is empty. Now that you know how to repeat elements inside your template with the new control flow syntax, let's learn about the alternative for the *ngSwitch directive:

```
@switch (condition) {
  @case (caseA) { Case A. }
  @case (caseB) { Case B. }
  @default { Default case. }
}
```

Just as with the other control flow blocks, you can use the @switch block interchangeably with the *ngSwitch directive. The new Angular control flow also introduced a new concept: the @defer block.

The @defer block allows you to lazy-load components or native HTML elements inside your HTML templates. The @defer block can lazy load and display elements based on different triggers, such as when a condition is met, when the elements enter the viewport, when the users interact with the placeholder, or based on a timer. The @defer block can improve your performance because fewer components have to be loaded when the user lands on a page. Additionally, the @defer block reduces the bundle sizes as lazy-loaded page elements don't have to be included in the initial application bundle. Here's an example of the @defer block with a placeholder:

```
@defer (on viewport) {
  <calendar-cmp />
} @placeholder { <div>Calendar placeholder</div> }
```

Now that you know about the new control flow syntax, let's explore the remainder of the features that were introduced by the Angular framework.

Other noteworthy new Angular features

Before we move on to the next section of this chapter, we will briefly go over other noteworthy improvements that have been made to the Angular framework:

- **Types forms**: In Angular 14, reactive forms were made fully type-safe. The values inside form controls, groups, and arrays are now type-safe across the entire API of reactive forms, enabling safer forms.

- **Improved page titles**: Since Angular 14, you can add a `title` property to your route configurations. This `title` property will set the page title without other implementations needed.

- **NgOptimizedImage**: `NgOptimizedImage` is a built-in image directive for optimized image fetching, rendering, and sizing. It has been stable for usage since Angular 15.

- **Functional approach**: Since Angular 15, you can use a functional approach for HTTP interceptors, route resolvers, and route guards.

- **Route parameter mapping**: This feature allows you to automatically map route data, query parameters, and path parameters to your component inputs. Because of this, you don't have to subscribe anymore, reducing complexity and boilerplate.

- **Injectable OnDestroy**: In Angular 16, the team introduced the `OnDestroy` injectable, which allows you to inject `DestroyRef` and access the `OnDestroy` life cycle hook more flexibly. It allows you to subscribe to the `OnDestroy` life cycle as well as inject it outside of your components.

- **Self-closing tags**: In Angular 16, you can use a self-closing tag for your component selectors. This can improve the readability of your HTML templates.

- **Required inputs**: In Angular 16, you can make component inputs *required*. If no input is provided in the template, the compiler will specify an error.

There are more newly added features, such as Vite support and better page hydration, but the ones we mentioned in this section are the most important for your daily development practices.

Now that you know about the new features that have been added to the Angular framework, we will move on and do deep dives into specific topics, starting with the Angular router.

A deep dive into Angular routing

This section will teach you about the **Angular router**, a powerful tool that handles navigation in your Angular applications. The router is responsible for seamless transitions without full page loads, updating the browser URL, as well as handling route data, redirects, query parameters, path parameters, route resolvers, and guarding routes from unauthorized visitors.

Let's start by creating two new components we can navigate to.

Creating new components

We are going to use an Nx generator for this. You can write a custom generator, but I will use the built-in component generator for now. Right-click on the app folder inside your `expenses-registration` folder and select the Nx console. In the dropdown, search for `component` and select **@nx/angular - component**.

Follow these steps to generate the necessary components:

1. In the **name*** input field, enter `pages/expenses-overview-page`.
2. Check the **standalone** checkbox.
3. Click **Show all options**.
4. In the **changeDetection** select box, select **OnPush**.

In the top-right corner, click **Generate**.

Once you've completed these steps, repeat the same steps for creating the second component. You will only change the name to `pages/expenses-approval-page`.

Now, let's serve the *finance-expenses-registration application* with this command:

```
nx serve finance-expenses-registration
```

You can also use the Nx console to serve your application. Just select the application under the **PROJECTS** tab and click on the **play** button after hovering over **serve**.

When you open the application at `http://localhost:4200/`, you'll see a white screen. This is because you only have a router outlet in your `app.component.html` file, which displays the current route, and no routes have been configured for our application yet.

Your application is running, and two components are ready to route to, so let's configure some routes for your application.

Configuring routes in Angular applications

Before **standalone components**, you added routing for your application in your ngModules by importing `RouterModule` and providing it with routes in a `forRoot` or `forChild` method. Because we're using the latest Angular techniques, we won't use ngModules. When bootstrapping with a standalone component, your routing is configured differently. When you open the `main.ts` file, you'll see an `appConfig` object being passed to the `bootstrapApplication` function. Open your `app.config.ts` file to locate this `appConfig` object. Inside, you'll find your routing configurations:

```
provideRouter(appRoutes, withEnabledBlockingInitialNavigation())
```

Routing is configured by adding the `provideRouter` function inside the `providers` array of the `ApplicationConfig` object. When Nx created the application, it already set this up for us.

Inside the `provideRouter` function, you'll find an array of `Route` objects and a `with EnabledBlockingInitialNavigation` function, which is required for routing with server-side rendering. We aren't using server-side rendering, so you can delete `with EnabledBlockingInitialNavigation`.

Open your `app.routes.ts` file to set up the routes for your application. To start, we will add two `Route` objects inside the `appRoutes` array – one for the expenses approval page and another for the expenses overview page:

```
export const appRoutes: Route[] = [{ path: 'expenses-overview',
component: ExpensesOverviewPageComponent },
{ path: 'expenses-approval', component: ExpensesApprovalPageComponent
}];
```

As you can see, each object has two properties: a `path` property to define the URL path and a `component` property for specifying the loaded component when we reach the path. For `ExpensesOverviewPageComponent`, we configured `expenses-overview`, which means it's accessible at `http://localhost:4200/expenses-overview`.

When you navigate to this URL, you'll see **expenses-overview-page works!** You can navigate to your routes inside TypeScript files by injecting `Router` as a dependency. Then, you can use this syntax:

```
this.router.navigate(['expenses-overview']); //Option 1
this.router.navigateByUrl('/expenses-overview'); //Option 2
```

You can navigate instead your HTML templates with `routerLink`, like this:

```
<a [routerLink]="['path', { outlets: { sidebar: 'path'} }]">Click to
navigate</a>
```

Now that you've configured two routes, let's explore what else you can configure in your route configurations.

Route configuration options

In this section, you will make your routing more robust and explore properties you can configure on your route objects.

Adding page titles

The `title` property was added in Angular 14 and is used to set HTML page titles dynamically. Before Angular 14, you needed subscriptions and a lot of logic to set the page title. With the introduction of the `title` property, Angular handles all of that behind the scenes and sets the page title for you.

You can set your titles with a simple string or `ResolveFn<T>`. When you use a string, you can use the following syntax:

```
{ path: '', component: ExpensesApprovalPageComponent, title: 'Expenses
Approval Page' }
```

We will use `ResolveFn<T>` to set page titles dynamically, like so:

```
export const titleResolver: ResolveFn<string> =
  (route: ActivatedRouteSnapshot) =>
    route.routeConfig?.path?.replace('-', ' ') ?? '';
```

This is a simple example where we take the route path and replace the hyphen with a space, but you can add any logic you want. Once you've defined your title resolver, you can assign it to your route configuration like this:

```
{ path: '', component: SomeComponent, title: titleResolver}
```

You can also overwrite the default behavior Angular uses to add the titles to your pages by overwriting the `TitleStrategy` class. This is only useful in edge cases, but it's good to know it's possible. We won't cover an example in this book, but you can find a simple overwrite of `TitleStrategy` in this book's GitHub repository: `https://github.com/PacktPublishing/Effective-Angular/blob/main/apps/finance/expenses-registration/src/app/app.routes.ts`.

Lazy loading standalone components

Lazy loading components is one of the most essential features of the router. It allows you to divide your application into small chunks that are only loaded when a user reaches a specific route of your application. With the introduction of standalone components, you can lazy load components, whereas before, you could only lazy load modules. Now, splitting your application into small bundles is easier because you can lazy load components instead of modules. You can lazy load each route by using the `loadComponent` property in your route configuration.

Since Angular 15, the router also supports automated unwraps of default imports. With automated unwraps, you don't need the `.then()` method to be chained to unwrap your import for the router. Because of this, we can make the syntax for lazy loading shorter and easier. Change your routes to lazy-loaded routes by changing the export of your component classes to `default` exports, like this:

```
export default class ExpensesApprovalPageComponent {}
```

Once you've done this, you can configure lazy-loaded routes like this:

```
{ path: '......', loadComponent: () => import('./pages/expenses-approval
page/expenses-approval-page.component') }
```

Do the same for your other route so that all your routes are lazy-loaded.

Next, we will learn how to use multiple router outlets and auxiliary routes.

Router outlets and auxiliary routes

If you want to develop dynamic user interfaces while reducing your bundle sizes even more, **named router outlets** are a great way to achieve this. We won't implement named router outlets, but I do want to explain how they work.

With named router outlets, you can lazy load specific sections of your pages and lazy load different components for these page sections based on your application state. You can, for example, display different sidebars on each page or even on the same page based on application states. Because you can lazy load these sidebar components, they will not be part of your main application bundle and will only be loaded when displayed. To use named router outlets, you need to configure a route with an `outlet` property, like this:

```
{ path: 'list', component: SomeComponet, outlet: 'sidebar'}
```

If you want to use lazy loading for these routes, you can replace the `component` property with the `loadComponent` property and import the component.

Routes with the `outlet` property defined can only be loaded by a router outlet with the same name specified on it. You can define named router outlets by adding a name attribute on the router outlet HTML tag like this:

```
<router-outlet name="sidebar"><router-outlet/>
```

With the named router outlet in your template and a route configuration with the `outlet` property defined, you have everything set up. Your main router outlet will work as expected and navigate to the expenses overview page when you add `/expenses-overview` after your root URL. The named router outlets work differently. The routes that are used by your named router outlets are called *auxiliary routes* and can be seen as sub-routes that operate independently from your main route. These auxiliary routes form a special kind of URL that looks like this: `http://localhost:4200/expenses-overview(sidebar:list)`.

As you can see, round brackets are added to your URL to represent your auxiliary routes. There is only one auxiliary route in our example, but there could be more, and they would all be inside the round brackets separated by a double forward slash. Your auxiliary routes are isolated inside these round brackets so that you can activate different auxiliary routes for the same main route.

Routing to auxiliary routes inside your TypeScript files can be done like this:

```
this.router.navigate(['path', { outlets: { sidebar: 'path'} }]);
```

When using `routerLink` in your HTML templates, you must add the following syntax to your HTML tag:

```
[routerLink]="['path', { outlets: { sidebar: 'path'} }]"
```

Now that you know about named router outlets, let's learn about route guards.

Route guards

The `canActivate`, `canMatch`, `canActivateChild`, and `canDeactivate` properties declare **route guards** in your route configurations. Route guards help you to secure routes and prevent users from accessing a route they are not intended to access. All four properties define a type of route guard that prevents the user from performing a specific routing task, such as activating or deactivating a route.

The implementations with the rules when these guards should allow or block a user are created by yourself and can contain any logic you need. Each route can configure multiple guard types, and you can add various implementations for each type. You can configure these guard types and the implementations for them in your route configurations, like this:

```
{ path: '......', loadComponent: ......,
    canActivate: [IsLoggedInGuard, IsAdminGuard],
    canDeactivate: [hasDoneSomeTaskGuard] },
```

In *Chapter 9*, we will create route guards and look at their implementations; for now, you just need to know that you can protect routes with route guards.

Now that you know about route guards, let's move on and start learning about child routes.

Defining child routes

Route configurations can also define **child routes**, which helps organize your routes better and easily create an initiative URL structure. Child routes are defined like this:

```
{ path: 'dashboard', component: DashboardComponent,
  children: [{ path: 'summary', component: SummaryComponent }]}
```

The preceding example would load `SummaryComponent` on the `/dashboard/summary` route. You can configure the same route without using child routes, but using child routes offers some advantages. The most apparent benefit is that you can better organize your routes. Another advantage of child routes is that you can share route resolvers and guards. When you use a route guard on a parent route, the guard will automatically be applied to all child routes. You can also use child routes to omit the round brackets in the URLs of your auxiliary routes from the named router outlets. However, there are some drawbacks to this compared to regular auxiliary routes. When using child routes to omit the round brackets, you can't load different auxiliary routes on the same main URL; instead, you need to add a new configuration for each route and auxiliary route combination.

Fallback routes and redirecting

You can configure **fallback routes** by using a double asterisk for the path. Your fallback route will be triggered when no route to the current browser URL is found. Most of the time, the fallback route is used to display a **404 Page Not Found** error. You can configure fallback routes like this:

```
{ path: '**', component: NotFoundComponent }
```

When working with child routes and named router outlets, you can configure multiple fallback routes, but in most scenarios, one fallback that redirects to a **404 Page Not Found** page will be enough. Another useful route configuration is a **redirect route**. You can use redirects to send a user to a specific route when some other route is loaded. When working with child and auxiliary routes, redirects are especially useful because often, you want to redirect to a specific child or auxiliary route when a parent or root route is accessed by the user. Because we don't have any auxiliary or child routes, we will add a simple redirect and send the user to the /expenses-overview route when they load the root route. You can add this redirect to your appRoutes array like so:

```
{ path: '', pathMatch: 'full', redirectTo: '/expenses-overview' }
```

Now that you know about fallback routes, let's dive into route resolvers.

Route resolvers

Route resolvers can resolve data before a route is activated and provide that data to your component. That might sound nice, but your route won't be activated until the data is fetched and can be passed to the route. As a result, when you fetch asynchronous data and the API isn't responding, the route will not be activated, and the user will be staring at a white screen. Resolvers should only be used if you have some edge case where a component cannot work without having specific data before the component renders. A simple implementation of a route resolver function looks like this:

```
export const userDataResolver: ResolveFn<User> = (
  route: ActivatedRouteSnapshot) => inject(UserService)
    .getUserData(route.paramMap.get("userId")).catch(......);
```

You can declare the route resolver on your route configurations like this:

```
{ path: 'path', resolve: productResolver, component: ......},
```

You can access the resolved data inside your components using the data property of the route snapshot:

```
protected readonly route = inject(ActivatedRoute)
ngOnInit() { this.route.snapshot.data; }
```

Don't use route resolvers unless you don't have any other option. Such scenarios don't arise often, if at all; however, I wanted to mention resolvers and make you aware of them and their drawbacks. When working on Angular applications, you will find route resolvers quite often in the code base.

Now that you've created components, set up routes, and learned about the Angular router, we will learn about component communication.

Component communication

This section will dive deep into **component communication**, starting with input and output decorators. Before we begin, let's create a new component with the Nx generator so that we have something to work with.

Name your new component `navbar` and add it to the `shared-ui-common-components` library. Don't forget to check the `standalone` checkbox and select `OnPush` for `changeDetection`. When the component has been created, add it to the `index.ts` area of your library:

```
export * from './lib/navbar/navbar.component';
```

After that, add the `navbar` component to the `app.component.html` file of your `expenses-registration` application. It's important to note that you need to add the `NavBarComponent` class to the `imports` array of your app component decorator. This is because we are using standalone components, and a standalone component needs to import everything it uses. Once you've added the `navbar` component to the template of your app component, you can get the code for the HTML and SCSS of the navbar from this book's GitHub repository: `https://github.com/PacktPublishing/Effective-Angular`.

Because `navbar` is also a standalone component, you need to add `RouterLink` and `CommonModule` to the `imports` array of the component decorator. These two imports are necessary because we use the `routerLink` and `*ngFor` directives in the template of the `navbar` component.

Now that you've created and added the `navbar` component to the `app` component template, we can look into parent-child component communication.

Receiving values with the @input() decorator

As we explained in *Chapter 1*, when we develop Angular applications, we divide our components into smart and dumb components. Dumb components are presentational components that are used in the templates of smart components. These dumb components should only receive data through `@Input()` decorators (alternatively, you can use the new `input()` Signal that was introduced in Angular 17; we will dive deeper into Signals in *Chapter 7*, so for now, we will use the decorator); dumb components do not inject services for data as that is the responsibility of smart components.

`@Input()` decorators are only defined on child components; the parent components pass data to the input. A component can be considered a child component when it's declared inside another component's HTML template. On the other hand, the component that declares a component in its HTML template is regarded as the parent component. Dumb components are always meant to be child components, whereas smart components can be both. Still, smart components are generally used as parent components and seldom declare input and output decorators.

Our newly created `navbar` component is a dumb component that's used as a building block for our pages. Since it's a dumb component, it must rely on input decorators to receive its data. A `navbar` component needs `navbar` items, so let's define an interface and input. First, define the interface in a new file or underneath your `navbar` component's class:

```
export interface NavbarItem {label: string; route: string;}
```

Here, we defined the interface. Now, let's add the input to the `navbar` component, like this:

```
@Input() navbarItems: NavbarItem[] = [];
```

Here, `@Input ()` is a decorator for the `navbarItems` field that tells the Angular compiler that the property can receive input from parent components. We gave the field a `NavbarItem` array type and an empty array's default value. If you don't give it a default value, the compiler will start to complain; you can prevent this by adding an exclamation mark after the property's name, like this:

```
@Input() navbarItems!: NavbarItem[];
```

In our example, we declared the `navbar` component in the `app` component template from our *expenses-registration application*, making `navbar` a child component of the `app` component. To pass our new input property data, let's declare a `NavbarItem` array inside the `app` component class, like this:

```
navItems: NavbarItem[] = [{ label: 'home', route: '/' }, { label:
'expenses approval', route: '/expenses-approval'}];
```

We want to pass this `navItems` array to the `navbar` component using the `navbarItems` input property. We can do this in the HTML template of the `app` component, where we declare the HTML selector tag for the `navbar` component. You can pass the `navItems` array as input using this syntax:

```
<bt-libs-navbar [navbarItems]="navItems" />
```

On the left-hand side, between the square brackets, you'll use the property name of the input property that's declared inside the `navbar` component – in our case `navbarItems`. On the right-hand side, you must assign the input with a value that's declared in the parent component – in our case, the `navItems` array.

It is important to know that when a component receives input values, the `ngOnChanges` life cycle hook is triggered – once this component has been created, before the `ngOnInit` life cycle hook runs, and then once more every time an input receives a new value. You can access the previous and new input values inside the `ngOnChanges` method like this:

```
ngOnChanges(changes: SimpleChanges) {console.log(changes)};
```

The current value inside the `SimpleChanges` object should be equal to the value of the input property declared in the component. So, if you need the current value, you can also access the component property. In our example, this would be `navbarItems`.

The `ngOnChanges` life cycle hook is a good place to perform extra logic when you receive an input value. Still, this can become messy when you have a lot of input properties and want to perform logic for each when they receive a new value. If this is the case, or if you want to transform the value into something else, you can use the `@Input()` decorator as a getter and setter.

Let's say we want to add `NavbarItem` for the home page to our input each time it receives values so that we don't have to declare the home page and its route inside the object we pass as input to the `navbar` component. We can do this by transforming the input into an input with a getter and setter. Start by adding a private property to the `navbar` component, like this:

```
private _navItems: NavbarItem[] = [];
```

Now, change the `navbarItems` input property to a getter and setter and use the private property inside the getter and setter, like this:

```
@Input()
set navbarItems(value: NavbarItem[]) {
   this._navItems = [{label: 'home', route: '/'}, ...value];
}
get navbarItems(): NavbarItem[] { return this._navItems; }
```

When the input property receives new input values, it will set the private property and add the home page route to these values. When you use the `navbarItems` property inside the `navbar` component or template, it will use the getter, which returns the private property, including the extra home page item. After you change the input property, you can remove the object for the home navbar item from the `navItems` array declared in your `app` component.

Since Angular 16.1, you can achieve the same with the **transform property** of the `@Input` decorator instead of creating a getter and setter. Using the `transform` property requires a lot less code and looks much cleaner. Let's convert our getter and setter into the `transform` property. First, remove the private `_navItems` property and the getter and setter we just added and replace them with this:

```
@Input({transform: addHome}) navbarItems: NavbarItem[] = []
```

Now, you only need to add a function called `addHome` with the transformation logic. You can add this function to a separate file or the same file underneath the `navbar` component class. The function looks like this:

```
function addHome(items: NavbarItem[]) {
   return [{ label: 'home', route: '/' }, ...items];
}
```

That's all you need; no more private property or getter and setter needed for transforming input values! If you need to perform other logic, such as setting component properties or running functions inside the component class, you can still use the getter and setter approach.

Lastly, since Angular 16, you can also make input properties required. When you make an input property required, it needs to be declared on the HTML tag when the component is used in a component's template. The parent that's using the component in its template must add the input property to the HTML tags and pass it a valid value; otherwise, the compiler will throw an error. To make our navbarItem input property required, we can change the input decorator like this:

```
@Input({ transform: addHome, required: true }) navbarItems:
NavbarItem[] = [];
```

Now that you know how to input values into a component, transform these inputs, or perform extra logic when a component receives new input values, we will learn how to emit values using the @Output() decorator.

Emitting values with the @Output() decorator

Child components also need a way to send events and data to the parent component. For example, if you have a table component in which you can display and update data, the table component shouldn't inject services to receive and update the data. This would result in a tight coupling of the table component and the data it displays. Each time you use the table with different data, it needs to add extra services and new logic to persist the data updates, and this is not a desirable situation.

Instead, the table component should be a dumb component that receives data as inputs and emits an event with the updated data as output. By doing so, your table component remains reusable and doesn't create unnecessary dependencies. The parent components are smart components that are used for specific business use cases or pages, so each can implement whatever logic is needed to handle the data updates for its specific page or business use case without creating unwanted dependencies.

To emit an event to a parent component, we need to create something that can emit our events. We can do this with the following syntax:

```
@Output() dataChanged = new EventEmitter<tableData>();
```

On the left-hand side, we have the @output decorator and the property name for our EventEmitter. On the right-hand side, we assign the property with new EventEmitter, and between the arrow symbols we add the type we wish to emit – in this example, this is tableData.

Next, you need to listen for the dataChanged event in the HTML template of a parent component where the data table component would be defined. Listening for @Output works the same as listening for regular DOM events such as *click* and *mouseleave*:

```
<bt-lib-table (dataChanged)="handleDataChange($event)" />
```

Between the round brackets, you can define the event's name, similar to `click`; in our example, we named the event `dataChanged`. On the right-hand side, you call a function you've created in the component class of the parent component. So, `$event` will contain whatever values you emit from the child component.

Lastly, we must *emit* events from the child component using the `dataChanged` property. Inside the `table` component, whenever the data changes and you want to emit an event to the parent, you can use the following syntax:

```
this.dataChanged.emit(updatedTableData);
```

As you can see, you can use the `dataChanged` property and call the `.emit()` method on it. You can pass whatever value you want to emit to the parent component inside the brackets. In this example, that would be the updated table data. To better illustrate how the input and output mechanism works, *Figure 2.1* shows how data flows from the parent to the child and vice versa:

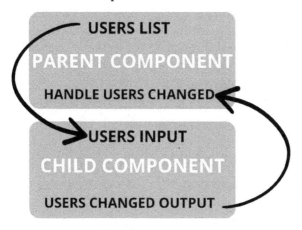

Figure 2.1: Hierarchical dependency creation

Now that you also know how to emit events to the parent component using the `@output` decorator, let's explore how we can combine the input and output decorators to create custom two-way data bindings.

Two-way data binding using @Input and @Output

Two-way data binding lets your components listen for events and update the corresponding value simultaneously. Let's say we have a custom `select` drop-down component with two input properties, one for the select options and one for the selected option. Besides this, the component also has a `selectionChanged` output to emit a value when a new selection is made.

In the parent component, we have a property for the currently selected value of the select component. This property is used as an input for the selected input and needs to be updated whenever a new value is selected and the `selectionChanged` output is emitted. To achieve this, you must have something like the following inside the template of the parent component:

```
<select [selected]= "selectedValue" (selectionChanged)="this.
selectedValue = $event" />
```

We can improve the preceding code snippet with two-way data binding and the banana-in-a-box syntax. For two-way data binding to work, the input and output properties in the child component must have the same name; only the output needs to add `changed` after the name. So, in our example, the input is named `selected`, which means the output needs to be named `selectedChanged`. When we use this naming convention, Angular knows to handle it as two-way data binding.

To use two-way data binding, in the parent component, we must update the HTML like this:

```
<select [(selected)]= "selectedValue" />
```

That looks a lot cleaner! With the preceding syntax, we pass the `selectedValue` property of the parent component as an input to the selected input property of the child component. When the child component emits the `selectedChanged` event, the `selectedValue` property will automatically be updated in the parent component. Take a look at the following syntax:

```
<select [(selected)]= "selectedValue" />
```

This is the same as using the following:

```
<select [selected]= "selectedValue" (selectedChanged)="this.
selectedValue = $event" />
```

As you can see, the banana-in-a-box syntax combined with the square and round brackets is much cleaner and more compact. Alternatively, you can use the new `model()` Signal that was introduced as a developer preview in Angular 17.2, but we will cover this scenario in *Chapter 7*.

Now that you know how to input, output, and two-way bind properties with the input and output properties, let's look for another way to communicate data between components and routes.

Other component communication methods

There are a few other means of communication for Angular components. You can access public properties and methods of child components with the `@ViewChild` decorator, communicate with child, parent, sibling, and unconnected components with *services*, and pass data in various manners to components with the *router*. Let's start with the `@ViewChild` decorator.

Using the @ViewChild decorator to access child components

The @ViewChild decorator is used to access template elements inside the component class. As in alternative to the decorator, you can also use the new viewChild() Signal that was introduced in Angular 17.2; we will cover this in more detail in *Chapter 7*. Using the @ViewChild decorator to access or update child properties and methods is straightforward but has some drawbacks. When using @ViewChild to communicate with your components, you can mutate values within your child component, which can lead to unexpected behavior and bugs that are hard to debug. Besides that, it makes your component hard to test. If you have a scenario where you need to update properties from the child in the parent, here's the syntax:

```
@ViewChild(NavbarComponent) navBar!: NavbarComponent;
```

Here, you declare the decorator; inside the function brackets of the decorator, you enter the child component's class name, then give it a variable name and type it with the component's class name. After the view of the parent component has been initialized, you can access the child component and its public properties and methods like this:

```
this.navBar.navbarItems;
```

As mentioned previously, I'm not a fan of this decorator, and it's recommended not to use it unless you need to achieve something that can't be done in another way.

Now that you know how to use the @ViewChild decorator to access properties and methods in child components, let's explore communication through the Angular router.

Component communication with the Angular router

The *router* is meant to navigate between routes but can also send data to a component that's been loaded on a route. The most common examples are route parameters and additional query parameters in the route, but you can also add data to the route using the data property in the route configuration or with route resolvers.

Let's say we have the following route configuration:

```
const routes = [{ path: 'dashboard/:id',
    component: DashboardComponent,
    data: {caption: 'Dashboard caption'},
    resolve: {permissions: DashboardResolver}}]
```

We also have this URL in our browser: https://some-url.com/dashboard/123?queryParam=paramValue.

When we reach this route, the `dashboard` component will be loaded. In that component, we can access the dashboard ID, the caption we added to the `data` property, the resolved permission data, and the value of the `query` parameter. To do so, you need to inject `ActivatedRoute` into the constructor or use the `inject` function:

```
private route = inject(ActivatedRoute);
```

After that, you can access the properties in the route snapshot like this:

```
this.route.snapshot.paramMap.get('id');
this.route.snapshot.queryParamMap.get('queryParam');
this.route.snapshot.data['caption'];
this.route.snapshot.data['permissions'];
```

You can also access the properties more reactively by subscribing – just remove the `snapshot` property and then add your subscription logic instead of accessing the properties through the `get` method:

```
this.route.queryParamMap.subscribe(......)
```

You can subscribe to the router's `paramMap`, `queryParamMap`, and `data` objects. Since Angular 16, you can also directly bind the route values to your component's `@Input()` decorators. To achieve this, you need to add `withComponentInputBinding` to your app config where you provide your routes:

```
provideRouter(appRoutes, withComponentInputBinding()),
```

Next, you can define the inputs in your `component` class, like this:

```
@Input() caption?: string;
@Input() id?: string;
@Input() queryParam?: string;
@Input() permissions?: string;
```

If you want to use a different property name, you can alias the inputs like this:

```
@Input('caption') captionFromRouteData?: string;
```

This will probably give you a linting error because aliasing inputs is not recommended, but when using input aliases for binding your route data, it's not a bad practice. So, in this case, you can disable the linting rule if it's throwing an error.

Now that you know everything about component communication using decorators and route data, we will start to look at dependency injection, which can be used to provide data to every part of your application.

Dependency injection

Dependency injection is one of the cornerstones of the Angular framework. Angular relies on the dependency injection design pattern to manage the relationships and interactions between different elements within an application. When using dependency injection, there are two main roles: the **providers** and **consumers** of the dependencies. As such, Angular dependency injection lets you provide logic or data to sections of your application that need to consume the given logic or data. Dependency injection is vital for developing loosely coupled applications with more modular, testable, and maintainable code. Generally speaking, dependencies in an Angular application are services (classes decorated with the `@Injectable()` decorator), but they can be strings, functions, or anything else you want to provide throughout your application.

The `Injector` abstraction is the core element of the Angular dependency injection system and facilitates the connection between providers and consumers of your dependencies. Make sure you distinguish `Injector` from the `@Injectable()` decorator, which marks a class as a candidate for dependency injection. The `Injector` abstraction checks if an instance of a dependency has already been created and provides it if the dependency has already been registered; it provides the dependency and registers if it hasn't been registered before. When your application is bootstrapped, Angular creates an application-wide *root* `Injector` and makes other injectors as needed for all dependencies that aren't accessible throughout the application.

Providing dependencies

You can provide classes and other values such as strings, dates, objects, and functions as dependencies. Both are provided differently. We will start with the most common use case, which is providing classes.

Providing classes as dependencies

When you provide a class in Angular, it's most likely an Angular service, but you can provide any class through dependency injection. The most common way to prepare a class for dependency injection is to decorate it with the `@Injectable()` decorator, like this:

```
@Injectable()
class SampleService {}
```

The @Injectable() decorator shows a class is used for dependency injection; you can do this by marking it as a root injector or by adding it inside a providers array of a specific component or ngModule. The @Injectable() decorator also ensures Angular can perform optimizations such as tree shaking. You can also provide classes without the @Injectable() decorator inside your providers array, but it's a good practice to mark them with the decorator unless you need to provide the classes in a specific place. Here's an example of how to provide a service inside the providers array of a component:

```
@Component({........ , providers: [SampleService] })
class ListComponent {}
```

When using the component-level providers array, the provided dependency becomes available for dependency injection for each component instance and all child components or directives used within the component's tree. If you add a dependency inside the providers array of an ngModule, the dependency can be injected and accessed everywhere within that module. You can declare classes inside the providers array of your modules like this:

```
@NgModule({ declarations: [ListComponent],
    providers: [SampleService]})
class AppModule {}
```

In our *expenses-registration application*, we don't have ngModules. When developing an Angular application without ngModules, you can provide dependencies inside your ApplicationConfig object; this is similar to marking a dependency as a root injector because it will be available for dependency injection throughout your application. Adding classes to your providers array inside the ApplicationConfig objects works like this:

```
export const appConfig: ApplicationConfig = {
    providers: [SampleService],
};
```

Lastly, the most common way to provide your classes as dependencies is by marking the class as a *root* injector. You can do this inside the @Injectable() decorator with the providedIn property, like this:

```
@Injectable({ providedIn: 'root' })
```

Now that you know how to provide classes as dependencies with the @Injectable() decorator, let's learn how to provide classes and other values such as strings, Booleans, and functions using provider objects.

Providing dependencies with provider objects

When you provide dependencies with a **provider object**, you always declare them inside a `providers` array. You can use two properties to declare a dependency with a provider object:

- **Provide**: The `provide` property holds the **provider token** that's used to identify and inject your dependencies. It's also used to configure the `Injector` instance.

- **Provider definition**: The second property is used to tell `Injector` how to create the dependency, and it can be defined with four values:

  - `useClass`: This tells Angular to provide the given class when the corresponding provider token is used

  - `useExisting`: This aliases another provider token and accesses the same dependency with two different tokens

  - `useFactory`: This defines a factory function to construct the dependencies based on some logic

  - `useValue`: This provides a static value such as a string or date as a dependency

Now, let's explore the four provider definitions in more detail.

Declaring provider objects with useClass

Here's an example of a provider object using the `useClass` provider definition:

```
providers: [{ provide: Logger, useClass: Logger }]
```

The preceding syntax is the same as using the following syntax:

```
providers: [Logger]
```

In the scenario where you only supply a class name, the provider object is automatically created behind the scenes. The provider object uses the class for the provider token and definition, so when would you use the provider object instead of only using the class?

Commonly, `useClass` is used when you want to overwrite a dependency injection class with a new implementation. In the preceding example, we provided a `Logger` class; let's say you create a new `BetterLogger` class extending the original `Logger` class. If you have a large application and the `Logger` service is used throughout your application, it's a lot of work to change the service everywhere it's declared. Instead of updating all dependency injection consumers, you can create a provider object and return the `BetterLogger` class for the `Logger` token:

```
providers: [{ provide: Logger, useClass: BetterLogger }]
```

If you provide the `Logger` service inside the `providers` array as well, you must make sure that you declare the new provider object with the `BetterLogger` class below your previous provider object, or simply remove the old object. If the `Logger` service is provided using the `providedIn` property inside the injectable decorator, the overwrite with `BetterLogger` will just work without any gotchas.

Now that you know how to create provider objects with the `useClass` provider definition, let's examine the `useExisting` provider definition.

Declaring provider objects with useExisting

The `useExisting` property allows you to map one provider token to another provider token, making sure two tokens will return the same dependency:

```
providers: [ BetterLogger,
  { provide: Logger, useExisting: BetterLogger}]
```

In the preceding example, you can inject the better logger using the `BetterLogger` and `Logger` tokens. Both will provide the same instance of the service. Be careful not to use the `useClass` definition for this:

```
providers: [ BetterLogger,
  { provide: Logger, useClass: BetterLogger}]
```

You might think the preceding example gives the same result, but it creates two tokens called `Logger` and `BetterLogger` that each return an instance of the `BetterLogger` class instead of the same instance.

Now that you know about `useExisting`, let's explore `useFactory`.

Declaring provider objects with useFactory

The `useFactory` provider definition allows you to define a function as a dependency. This can be just a regular function you want to inject into multiple places of your application or a factory function that constructs a service class. Let's say you have `AdminDashboardService` and a regular `DashboardService` you want to inject using the `DashboardService` tokens, depending on the active user role. You can achieve this with `useFactory`. First, create your factory function:

```
const DashboardServiceFactory = (userService: UserService) =>
userService.user.isAdmin ? new AdminDashboardService() : new
DashboardService();
```

Next, declare the provider object inside your `providers` array, like this:

```
{ provide: DashboardService,
  useFactory: DashboardServiceFactory, deps: [UserService] }
```

The factory function relies on `UserService` to check if the current user is an admin user. Because of this, you need to add `UserService` inside a `deps` array of your provider object.

Now that you know more about `useFactory`, let's export `useValue`.

Declaring provider objects with useValue

The `useValue` provider definition is the simplest one. It returns a constant value such as a string, date, or Boolean. It's useful for providing things such as a base website URL, base API URLs, or other constant values:

```
{ provide: BASE_URL, useValue: 'www.someurl.com/'}
```

Now that you know about all four provider definitions, let's explore the `InjectionToken` object.

Using InjectionToken as a provider token

When you declare a provider object, you supply the `provide` property with a provider token. The provider token is used to inject the dependency into the consumers of the dependency. You can use three values for the provider token, but only two should be used. You can use a class name, `InjectionToken`, or a string. Only the class name and `InjectionToken` should be used. When providing a class-based dependency, you should use the class name as the provider token; when using a non-class-based dependency, you should use an `InjectionToken` object. You can create an `InjectionToken` object like this:

```
export const BASE_URL = new InjectionToken<string>('URL');
```

You can export a constant, give it a name – in our case, `BASE_URL` – and assign it with `new InjectionToken`. The value between the arrow brackets is the type of your dependency – in this example, a string – and the value between the round brackets is a description for your `InjectionToken`. Now, you can use this `InjectionToken` in a provider object like this:

```
{ provide: BASE_URL, useValue: 'www.someurl.com/'}
```

When assigning a provider token, you can also use a simple string like this:

```
{ provide: 'BASE_URL', useValue: 'www.someurl.com/'}
```

But there are some reasons why you should always use `InjectionToken` instead of a string:

- `InjectionToken` objects are type-safe and allow TypeScript to perform type-checking on your injected value. When you use a simple string, the compiler will not know what type your dependency is.

- When using a string, you can run into name collisions, meaning you can assign two dependencies with the same string, which can result in wrongly injected values, errors, and bugs. This doesn't have to be because you define two dependencies with the same string; it can also happen when a dependency you use from a third-party library uses the same string as a provider token. The `InjectionToken` object ensures a unique value is used for your provider token.

- When you minify your code during the production build, string values can be renamed. This can result in problems with your dependency injection system.

Now that you know how to provide dependencies and what `InjectionToken` objects are, let's learn how to inject and consume dependencies.

Injecting dependencies

There are two ways to inject dependencies: **constructor injection** or the new **inject function** introduced in Angular 14. At their core, both constructor injection and the inject function do the same thing, but with the `inject` function, you can improve some architectural patterns and inject values in places where you don't have a constructor. You can inject services everywhere within your Angular applications – in component classes, services, other classes, and even in functions you export. Let's examine how to inject class-based dependencies using constructor injection and the `inject` function, starting with constructor injection:

```
constructor(private logger: LoggerService) {}
```

Doing the same thing with the `inject` function looks like this:

```
private logger = inject(LoggerService);
```

You can use the two examples mentioned here interchangeably; both achieve the same thing, only with different syntax. When injecting class-based dependencies, you inject the dependency using the provider token, which is typically the class name. When you inject non-class-based dependencies, you need to use `InjectionToken` as a provider token. The syntax to do this is slightly different in the case of constructor injection. You need to use the `@Inject()` decorator function inside your constructor like this:

```
constructor(@Inject(BASE_URL) private baseUrl: string) {}
```

Injecting `InjectionToken`-based dependencies with the `inject` function can be done like this:

```
private logger = inject(LoggerService);
```

As you can see, you don't need the `@Inject()` decorator for the `inject` function; you can simply use `InjectionToken` inside the function brackets of the `inject` function, and the rest is done for you. After injecting a dependency, you can use it like any other value:

```
this.logger.logValue('Value to log');          this.baseUrl;
```

Now that you know how to inject dependencies, let's dive deeper into the `inject` function, why you should use it instead of constructor injection, and where to declare the `inject` function.

Using the inject function for better dependency injection

The `inject` function is more flexible than constructor injection because it can be used in more places and works better when using inheritance. You can declare the `inject` function anywhere, but it needs to run inside the *injection context*; otherwise, you'll get an error. The injection context is inside the constructor of a class, where you initialize fields inside a class inside the factory function of `useFactory`, route guards, route resolvers, and HTTP interceptors:

```
export class AppComponent {
   private url = inject(BASE_URL); // Is injection context
   constructor() { // Is injection context }
   someMethod() { // No injection context }
}
```

One of the main advantages of the `inject` function is that it allows you to abstract logic into a function. Let's say you want to fetch dashboards; you can abstract this logic into a separate function using the `inject` function. To do this, you can create a function like this:

```
export const fetchDashboards = (): Observable<Dashboard[]> =>
inject(HttpClient).get<Dashboard[]>('api/dashboards');
```

Here, we use an arrow function. Because we don't use brackets, the arrow function directly returns what we write after the arrow. This is the same as writing this:

```
export const fetchDashboards = (): Observable<Dashboard[]> => { return
inject(HttpClient).get………. }
```

Our function uses the `inject` function to inject the Angular `HttpClient` as a dependency to fetch the dashboards. You can use this `fetchDashboards` function inside the injection context of your components and services. For example, you can assign the function to a component property and subscribe to it in your template using the async pipe:

```
export default class DashboardsListComponent {
   dashboards$ = fetchDashboards();
}
```

Now, inside the template of this component, you can do something like this:

```
<h1 *ngFor="let dashboard of (dashboards$ | async)">
   {{dashboard.title}}
</h1>
```

When using the `fetchDashboards` function outside the injection context, such as in a method, you get an error telling you that the `inject` function can't be used outside the injection context. But there is a solution to this: you can use JavaScript closure to use the `fetchDashboards` function anywhere, even outside the injection context. To use closure, adjust the `fetchDashboards` function so that it returns a function:

```
export const fetchDashboards = (): Observable<Dashboard[]> => {
    const http = inject(HttpClient);
    return () => http.get.........
}
```

As you can see, we returned an arrow function, which returns the HTTP call instead of directly returning the HTTP call. Inside your `component` class, you must assign this closure function to a property, like this:

```
protected _fetchDashboards = fetchDashboards();
```

The `fetchDashboards` function returns another function, and because of that, the `_fetchDashboards` property is also a function. This `_fetchDashboards` function now holds the function that returns the observable HTTP call for fetching the dashboards and can be used anywhere in your class, also outside the injection context:

```
export default class AppComponent {
  private _fetchDashboards = fetchDashboards();
  loadDashboards() { this._fetchDashboards().subscribe(…) }
}
```

Before converting the `fetchDashboards` function so that it can use closure, it can't be used inside the `loadDashboards` function of the component because this is outside the injection context. By using a closure and returning the function to a property in our injection context, we can now use the function to fetch the dashboards outside the injection context. Using this pattern allows for great abstraction of logic and functions with a single responsibility that can be shared throughout your application while keeping component classes simple and clean.

Besides abstracting logic into dedicated functions, the `inject` function offers some advantages when working with inheritance. When using dependency injection in both the base class and the class inheriting from the base class, you need to call `super()` and pass the dependencies to the base class. Here's a simple example:

```
export class baseService {
  constructor(private router: Router) { }
}
export class DashboardService extends baseService {
  constructor(private logger: Logger, router: Router) {
```

```
    super(router);
  }}
```

Calling `super()` and passing along dependencies can become a hindrance and look messy. With the `inject` function, you can prevent this. By using the `inject` function in both classes, you don't have to pass along anything or call the `super()` method anymore:

```
export class baseService {private router = inject(Router);}
export class DashboardService extends baseService {
  private logger = inject(Logger);
}
```

As you can see, you don't have to pass along `router` to the base service, and there is no need for the `super()` call anymore; that looks much cleaner! Let's say you only used the router for navigation inside the base class. In that case, you can even take it a step further and abstract the navigation functionality into a separate closure function, as we did before, with the `fetchDashboards` function:

```
export const navigateFn = () => (url: string) => inject(Router).
navigate([url]);


export class baseService {
  protected _navigateFn = navigateFn();
}
```

Because we're using an arrow function that returns another arrow function, we can use `_navigateFn` outside the injection context. Inside the service that inherits the base class, you can navigate using the following code:

```
this._navigateFn('some/url');
```

Now that you know why, when, and how to use the `inject` function instead of constructor injection, let's explore how dependency injection creates instances of the injected services and what hierarchical injectors are.

Dependency instances, injector hierarchy, and resolution modifiers

The last thing you need to learn is how Angular creates instances of the dependencies you inject, how the **injector hierarchy** works, and how to control it with **resolution modifiers**.

Each time you inject a dependency, Angular will check how the dependency is provided, starting at the lowest level of the injector hierarchy – the `providers` array inside your component or directive. If Angular can't find a provider, it will start to move up in the injector hierarchy, first to the parent components; if it can't find the provider in any of the parent components, Angular will check the ngModule of the component or your `ApplicationConfig` object in case you don't use ngModules.

If Angular can't find a provider in the ngModules, it will look in the root injector for a service with the @Injectable decorator and providedIn root. If Angular still can't find a provider for the inject dependency, Angular will throw an error. If you inject a dependency in a service, Angular will skip the component hierarchy steps.

Let's say we want to inject LoggerService as a dependency. When Angular finds the provider for LoggerService, it will check if the injector already created an instance of LoggerService at the hierarchy level where the provider was located. If the injector already made LoggerService for the given hierarchy level, it will return the created instance. Otherwise, it will create one and then return it. For each hierarchy level, a singleton instance will be created and shared by all consumers on the lower hierarchical levels. If two or more sibling components declare a provider, a singleton will be created for each provider, and each component will use its own singleton of LoggerService and share it with all its subsequent child components and directives.

Figure 2.2 depicts how hierarchical dependencies are created and shared:

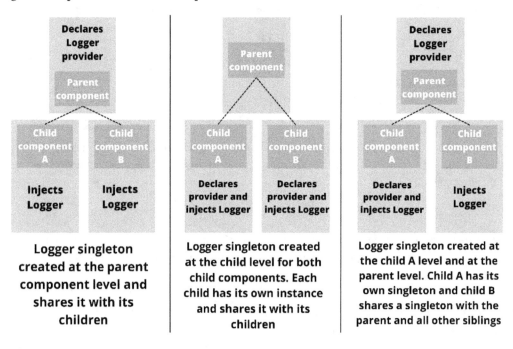

Figure 2.2: Hierarchical dependency creation

Knowing how Angular dependency injection creates and shares dependencies allows you to provide the correct instance to all application sections. If you want a singleton to be shared among all dependency consumers in the application, you just need to provide the dependency in the root injector. In some more complex and edge cases, you need to be able to have more control over how Angular looks for providers. In this scenario, you can adjust the behavior with resolution modifiers.

There are four resolution modifiers you can apply to your injected dependencies. Each has its own function and how you apply these modifiers differs depending on if you use constructor injection or the `inject` function. Let's examine these four resolution modifiers and what they do:

- `@Optional()`: The `@Optional()` modifier makes sure a dependency is optional. Angular won't throw an error if it can't find a provider for the given dependency.

- `@SkipSelf()`: With the `@SkipSelf()` modifier, you tell Angular it should start looking for a provider in the first parent in the dependency hierarchy.

- `@Self()`: The `@Self()` modifier tells Angular to only look at the component or directive itself for the provider.

- `@Host()`: The `@Host()` property marks a component as the last step in the search for a provider, even if there are more components higher up in the tree. When the `@Host()` property is found, Angular stops looking for the dependency provider.

Now that you know what resolution modifiers are available and what they do, let's see how you can apply them to dependencies. When you want to apply a resolution modifier to a dependency that's been injected with constructor injection, you can use the following syntax:

```
constructor(@Optional() @SkipSelf() logger?: Logger) {}
```

If you want to apply the resolution modifiers to a dependency that's been injected with the `inject` function, you can use the following syntax:

```
logger = inject(Logger, {optional: true, self: true});
```

Now that you know everything you need to know about Angular dependency injection, let's create a simple service before we move on to the next chapter.

To create this service, we are going to use the Nx generator again. First, create a new library called expenses under the `finance` domain with its type set to `data-access`. Next, open your Nx console again and click **Generate**. In the dropdown that appears, search for `service` and select @ **schematics/angular - service**.

Follow these steps to generate the service:

1. In the **name*** input field, enter `services/expenses`.

2. In the **project*** select box, select `finance-data-access-expenses`.

3. In the top-right corner, click **Generate**.

4. Go to the `index.ts` file of your *expenses library* and add the following code:

   ```
   export * from './lib/services/expenses.service';
   ```

That's all for now! With that, you've created a service and know everything you need to know about Angular dependency injection.

Summary

In this chapter, we explored some of the most fundamental features in the Angular framework and learned what Angular added in their latest major releases. You learned about Signals, the new control flow syntax, and `@defer` blocks. We covered how to communicate between components using input and output decorators, as well as less conventional methods for component communication using router data and the `viewChild` decorator. You also learned about the Angular router and how to configure route objects for more advanced scenarios and auxiliary routes. Finally, you learned about dependency injection in great detail. You now know what the difference is between constructor injection and dependency injection when using the new `inject` function. We also created some `provides` arrays and demonstrated how to declare the injection value and token. Lastly, you learned about the injector hierarchy and how you can control the provider that should be used by applying resolution modifiers.

In the next chapter, we will learn about Angular directives, pipes, and animations.

3

Enhancing Your Applications with Directives, Pipes, and Animations

When building frontend applications, we often need to enhance, transform, add, remove, or replace DOM elements and values within HTML templates. The Angular framework facilitates this through the use of directives, pipes, and animations. This chapter will explain how to create and use directives, pipes, and animations within Angular. By the end of this chapter, you will know all the ins and outs of directives, from directive composition to creating directives with powerful selectors. You will also learn about making custom pipes and using the built-in pipes effectively. Lastly, we will explore how to build and reuse animations within applications built using Angular.

This chapter will cover the following main topics:

- Using and creating Angular directives
- Transforming values using Angular pipes
- Creating and reusing stunning animations

Using and creating Angular directives

Directives come in two different types: **attribute directives** and **structural directives**. Angular has a list of built-in directives and allows you to create your own directive to cover your personal use cases. Since Angular 15, a new feature was introduced: **directive composition**. Directive composition allows you to assign directives inside component decorators instead of their template. Directive composition can also be used to declare directives inside the decorator of other directives, resulting in a directive that applies multiple directives simultaneously.

When you want to use a directive in a standalone component, you need to add the directive to the `imports` array of the component. If it's a built-in directive, you can also import `CommonModule`, as `CommonModule` contains all built-in directives. When you generate a component using the built-in Nx generator, `CommonModule` is added by default. This section will teach you everything about directives, starting with attribute directives.

Angular attribute directives

Attribute directives serve as tools for modifying DOM elements' attributes, behavior, or appearance. The directive makes modifications based on logic defined in a class decorated with a directive decorator. Attribute directives are assigned by adding the selector of the attribute directive to the HTML tag like this:

```
<div appRedBackgroundHover>Highlight me!</div>
```

In the preceding example, the selector of the directive selector is `appRedBackgroundHover`. When you use an attribute directive with an input, you use the square bracket syntax like this:

```
<div [ngClass]="useRedStyles ? 'red' : 'blue'">Click</div>
```

In the preceding example, `ngClass` is the attribute directive, and we add it between square brackets because it receives an input. Now that you know how to use attribute directives in HTML templates, let's explore the most commonly used built-in attribute directives and, after that, create a custom attribute directive.

Common built-in attribute directives

The most commonly used built-in attribute directives are the following:

- `ngClass` or `class`: These directives are used to conditionally add CSS classes.
- `ngStyle` or `style`: These directives are used to conditionally add inline styling.
- `ngModel`: This directive is used for two-way data binding on "from" elements.

More built-in attribute directives exist, but these additional built-in directives belong to specific Angular packages, such as `routerLink` from the **router** package and `formGroup` from the **forms** package. We've seen `routerLink` in *Chapter 2*, and we will learn more about directives from the forms package in *Chapter 4*. Now, let us learn how to create custom attribute directives.

Creating custom attribute directives

We will start by creating a new library with our custom generator. Name the new library common-directives and select **shared** for the domain and **ui** for the type. Next, we will use an Nx generator to create the directive for our new library. Go to libs\shared\ui\common-directives\src\lib folder, right-click, and select **Nx generate**. In the search bar, type directive and select **@nx/angular – directive**. Now, follow these steps:

1. Enter highlight for the **name*** field.
2. Click on **Show all options**.
3. Check the **standalone** checkbox.
4. Click on **generate** in the top right of the window.
5. When the directive is generated, export the directive in the index.ts file of the library:

```
export * from './lib/highlight.directive';
```

When you look at the highlight.directive.ts file, you will see an empty class with a directive decorator above it:

```
@Directive({
   selector: '[btLibsUiHighlight]',
   standalone: true,
}) export class HighlightDirective {}
```

Inside the decorator, you'll see the standalone flag set to true, indicating that the directive is standalone and doesn't have to be included in an NgModule. Besides that, you'll find the directive selector. The selector is used to apply the directive.

Now, let's add some logic in the HighlightDirective class so that our directive does something. Start by injecting ElementRef. ElementRef gives you access to the host DOM element, the HTML elements applying the directive. After injecting ElementRef, you need to use ElementRef to adjust the host element that applies the directive. Here is an example highlighting the host element with a background color:

```
export class HighlightDirective implements OnInit {
   private el = inject(ElementRef).nativeElement;
   ngOnInit() { this.el.style.backgroundColor = 'blue';}}
```

You can also inject ElementRef by using constructor injection instead of the inject function and then add the background color and text color inside the function brackets of the constructor. Alternatively, if you only want to set a style, CSS class, or attribute on the host element, you can use the @HostBinding() decorator:

```
@HostBinding('style.backgroundColor') get color() { return 'red'; }
```

Now, to apply this directive, you need to import it in a standalone component or NgModule and use the selector on an HTML element like this:

```
<div btLibsUiHighlight>I'm highlighted</div>
```

Let's improve the directive and make it configurable with @Input() decorators. The most common way to pass a value to a directive is to add an @Input() with the same name as the directive selector. In our case, this would look as follows:

```
@Input() btLibsUiHighlight!: string;
```

Now, you can use the btLibsUiHighlight property to assign the background color:

```
this.el.style.backgroundColor = this.btLibsUiHighlight;
```

To give the directive input a value, you need to add square brackets around the directive when you declare it on an HTML element and give it the value you want:

```
<div [btLibsUiHighlight]="'orange'">I'm highlighted</div>
```

If you don't like using the selector name in your TypeScript file but instead use a more descriptive property name, you can alias the input like this:

```
@Input('btLibsUiHighlight') background!: string;
```

This will probably give you a lint error because aliasing inputs is not generally recommended. Still, you can disable the lint error in this scenario if you prefer a more descriptive property name.

If you also want to customize the text color, add another @Input() property to the directive class. We will name the input property textColor and assign it a default value of white:

```
@Input() textColor = 'white';
```

This is how you assign additional inputs and their values in the HTML template:

```
<div [btLibsUiHighlight]="'orange'" [textColor]="'blue'"> I'm
Highlighted </div>
```

Declaring the btLibsUiHighlight input is required; the additional inputs are optional, but remember to give them a default value; otherwise, you might run into errors or unwanted UI behavior. You can make both inputs optional with default values; you have to rename the inputs so that no input has the same name as the directive selector:

```
@Input() background = 'black';
@Input() textColor = 'white';
```

Next, you change the directive on the HTML element by removing the square brackets and adding the additional inputs only if you want to overwrite the default value:

```
<div btLibsUiHighlight [background]="'orange'"
[textColor]="'blue'">I'm highlighted</div>
```

Lastly, we will handle user events in our directive. When handling user events in a directive, you often want to listen to events on the host elements. You listen to host element events by using the @HostListner() directive. Inside the function brackets of the @HostListner() directive, you need to add the browser event you want to listen for. This is how we can adjust our directive so it will apply the highlight for mouseenter and restore the original text and background color for mouseleave:

```
private el = inject(ElementRef).nativeElement;
private originalColor = 'black';
private originalBackground = 'white';

@HostListener('mouseenter') onMouseEnter() {
  this.originalColor = this.el.style.color;
  this.originalBackground = this.el.style.backgroundColor;
  this.el.style.backgroundColor = this.background;
  this.el.style.color = this.textColor;
}
@HostListener('mouseleave') onMouseLeave() {
  this.el.style.backgroundColor = this.originalBackground;
  this.el.style.color = this.originalColor;
}
```

We've added a property to save the original background and text color to restore that for mouseleave. Lastly, we've added the host listeners so we can react to the mouseenter and leave the events of the host element. We've added the logic to adjust the background and text color inside the host listeners.

Now that you know how to create custom attribute directives and enhance them with inputs and host listeners, let's explore structural directives.

Angular structural directives

Structural directives represent a tool for adding and removing DOM elements based on logic. The logic of when to add or remove DOM elements is defined in the directive class. Defining structural directives in your HTML templates is achieved like so:

```
<div *ngFor="let item of list">{{item}}</div>
```

In the preceding example, we applied the *ngFor directive; as you can see, it's prefixed with an asterisk. This asterisk is typical for structural directives and helps you to distinguish them from attribute directives in your HTML templates. Another difference between structural and attribute directives is that you can declare multiple attribute directives on a DOM element but only one structural directive. If you want to apply multiple structural directives, you need to wrap the DOM element with an ng-container tag and add the additional directive to the ng-container tag:

```
<ng-container *ngIf="expression">
  <div *ngFor="let item of list">{{item}}</div>
</ng-container>
```

Angular exposes built-in directives and allows you to create your own. First, let's look at some common built-in structural directives

Common built-in structural directives

The most commonly used built-in structural directives are the following:

- *ngIf: This is used to conditionally show or hide DOM elements (alternatively (to the directive), you can use the new @if control flow syntax, as shown in *Chapter 2*).

- *ngFor: This is used to create a for loop in your HTML template and output a DOM element for each item in an array (alternatively (to the directive), you can use the new @for control flow syntax, as shown in *Chapter 2*).

- *ngSwitch: This is used to create a switch case in your HTML template and display the DOM elements for the matching switch case (alternatively (to the directive), you can use the new @ switch control flow syntax, as shown in *Chapter 2*).

If you ever worked on an Angular application, you have most likely seen and used all of these structural directives before, as they are really common. Still, the *ngIf and *ngFor directives have some additional, lesser-known properties that I want to explain and showcase. After that, we will create a custom structural directive.

Using *ngIf to its fullest extent

The *ngIf directive is used to display DOM elements conditionally. The DOM elements do not render unless the property or statement is evaluated as true. Often, you need to display one block of HTML if the condition is true and another if it's false. A common solution to this is to use two *ngIf directives with an opposing statement like this:

```
<div *ngIf="showContent">Show if true</div>
<div *ngIf="!showContent">Show if false</div>
```

This is perfectly valid and very readable syntax, but `*ngIf` also allows you to create an `if-else` statement. You can do this with the following syntax:

```
<div *ngIf="showContent; else elseBlock">Show if true</div>
<ng-template #elseBlock>
  <div>Show if false</div>
</ng-template>
```

As you can see, this is a bit bulky in the HTML template because you need to use the `ng-template` tag for the content shown when the `else` statement is triggered. What syntax you want to use is up to you; there is no better or worse way; it's more about preference.

Using *ngFor effectively

When using `*ngFor`, you can add many properties to enhance the usage of the directive. The `*ngFor` directive outputs DOM elements for each item in a list. Often, when you output DOM elements for a list, you want to know the current index, if something is the first or last element, or if it's an odd or even index. Based on these values, you might want to add some styling classes or use specific properties in the template. The `*ngFor` directive allows you to detect these values like this:

```
<div *ngFor="let item of list;
    let i = index; let isFirst = first;
    let isLast = last; let isEven = even; let isOdd = odd"
>
  Item at index {{ i }}: {{ item }}
  Is first: {{ isFirst }} Is last: {{ isLast }}
  Is even: {{ isEven }} Is odd: {{ isOdd }}
</div>
```

As you can see, you can add variables to the HTML tag that defines the `*ngFor` directive to access values such as `index`, `first`, `last`, `odd`, and `even`. If our `list` would be the following array: `[0, 1, 2, 3]`, then the preceding code snippet would output the following result in the browser:

```
Item at index 0: 0 Is first: true Is last: false Is even: true Is odd:
false
Item at index 1: 1 Is first: false Is last: false Is even: false Is
odd: true
Item at index 2: 2 Is first: false Is last: false Is even: true Is
odd: false
Item at index 3: 3 Is first: false Is last: true Is even: false Is
odd: true
```

Besides the `index`, `first`, `last`, `odd`, and `even` properties, `*ngFor` also has something else to improve its performance.

By default, when you render something with *ngFor and something changes in the list, Angular will re-render the entire list. As you can imagine, this affects your performance negatively. You can add the trackBy function to improve this. When you use the trackBy function, Angular will identify each item by using the index or an ID. By doing so, it will only re-render things that change. It's recommended that you use the trackBy function as much as possible. In the template, you can define the trackBy function like this:

```
<div *ngFor="let item of users; trackBy: trackByFunction">
{{ item }} </div>
```

In your component class, you can define the trackBy function like this:

```
trackByFunction(index, user) { return user.id; }
```

Now that you know about the hidden features of *ngIf and *ngFor, let's see how you can create custom structural directives.

Creating custom structural directives

Creating your own structural directives is similar to creating attribute directives, but the directive class has some critical differences. In addition, the use cases are different. Custom attribute directives are useful for things such as auto-focusing on elements, applying different themes to specific elements, highlights, text resizing, tooltips, and popovers, as well as adding CSS classes, aria attributes, or IDs. Custom structural directives are used to remove or add DOM elements; some good use cases are an if false directive, a "repeat x number of times" function, and showing or hiding elements based on permissions or specific window sizes.

We will create a custom directive that shows an element when a condition is false, which is basically the opposite of the *ngIf directive. Start by generating the directive using the same steps we used for generating the custom attribute directive (see the *Creating custom attribute directives* section). The only difference will be the name; this time, name the directive ifFalse.

When your directive class is generated, you can start to add the logic for your structural directive. In the attribute directive, you injected ElementRef; for the structural directive, you must inject TemplateRef and ViewContainerRef. When you add a structural directive to an HTML element, Angular will convert it into an embedded template using the ng-template tags like this:

```
<ng-template [ ngIf ]="condition">
  <div>Shown when condition is true</div>
</ng-template>
```

The embedded template created by Angular is what you access with TemplateRef. The embedded template is not rendered unless the structural directives adds it to the view container using ViewContainerRef. ViewContainerRef gives you access to the view where the host element with the directive is defined.

Let's start by adding `TemplateRef` and `ViewContainerRef` to your directive class. You can inject `TemplateRef` and `ViewContainerRef` by using constructor injection or the `inject` function; I will use the `inject` function like this:

```
private templateRef = inject(TemplateRef);
private viewContainer = inject(ViewContainerRef);
```

We must also track if we already added the embedded view to the view container. To do this, add another private property to the directive class:

```
private embeddedTemplateAdded = false;
```

Next, we need an `@Input()` for our `if false` directive so we can give it a condition to assess. We will use an `@Input()` setter, so each time, it receives a new value; we can perform the logic to add or remove the embedded template to the view when the condition is evaluated. The `@Input()` setter needs to have the same name as the directive selector:

```
@Input() set btLibsUiIfFalse(condition: boolean) {}
```

Inside the `@Input()` setter, we will add the embedded template as an embedded view of the view container if the condition evaluates as `false` and if we have not already added it to the embedded template. In the case of an `if false` directive, when the condition evaluates to true, and considering we have already added the embedded template, we want to clear the view container so that the previously added embedded view is removed and Angular renders the original HTML again without the additional embedded template added to it. To achieve this, you can change `@Input()` as follows:

```
@Input() set btLibsUiIfFalse(condition: boolean) {
  if (!condition && !this.embeddedTemplateAdded) {
    this.viewContainer.createEmbeddedView(this.templateRef);
    this.embeddedTemplateAdded = true;
  } else if (condition && this.embeddedTemplateAdded) {
    this.viewContainer.clear();
    this.embeddedTemplateAdded = false;
}}
```

As you can see, we add `TemplateRef` as an embedded view when our check passes; otherwise, we clear the view container, given that we already added `TemplateRef`.

We can now use our custom structural directive like any other directive by using the directive selector with an asterisk in front of it:

```
<div *btLibsUiIfFalse="condition">shown when false</div>
```

Now that you know how to use built-in and create custom attribute and structural directives, let's learn what you can do with directive selectors.

Directive selectors

Directive selectors can be used to make your custom directives even more powerful. They allow you to automatically assign directives or limit the HTML elements that can use the directive. Let's take our custom `btLibsUiHighlight` directive and change the selector so that it will be applied to all span elements by default:

```
selector: 'span, [btLibsUiHighlight]',
```

When you use the preceding example as a selector, the directive will automatically be applied to all span elements, and you can add it to other elements using the `btLibsUiHighlight` selector. Now, let's say you need an option to exclude some span elements, so by default, all span elements receive the highlight directive, but when you want to opt-out, you can. To achieve this, you can add the `:not` syntax to the selector like this:

```
selector: 'span:not([noHighlight]), [btLibsUiHighlight]',
```

Now, all span elements will have the highlight directive applied unless you add `noHighlight` to a span element. For all other elements, you still need to add `btLibsUiHighlight` to apply the directive:

```
<span noHighlight>Test</span>
```

If you want to exclude HTML elements, you can achieve this using your selector by using the `:not` syntax like this:

```
selector: '[btLibsUiHighlight]:not(label)',
```

When you use the preceding selector, you can add the `btLibsUiHighlight` directive to all elements but the `label` element. When you try to add the directive to a `label` element, the compiler will throw up an error.

You can also make selectors that apply directives to HTML elements with a specific ID, data attribute, or CSS class applied to the HTML element. Here is an example for all three of these options:

```
selector: '#someId, .someCssClass, [data-highlight="true"]'
```

Now that you know everything you need to know about built-in directives, custom directives, and directive selectors, let's move on and learn about directive composition.

Angular directive composition

Directive composition is a relatively new concept that was introduced in Angular version 15. As the word indicates, directive composition lets you compose different directives on components and directives. It lets you declare directives in component and directive class decorators instead of adding them using HTML templates. You can use directive composition to automatically apply directives to components, just like directive selectors. Directive composition can also be used to create directives that apply multiple directives using a single selector.

Let's say we have a tag and button component and a type and size directive, allowing you to set a type (primary or secondary) and size (small, medium, or large), which applies a specific CSS class to the host element. If you want to automatically apply these two directives to all your buttons and tags, you can use directive composition to achieve this. Add a hostDirectives array inside the component decorator to add the directives to the component. In the hostDirectives array, you can add objects with the directive and the inputs for the decorator. If the decorator has no inputs, you can add the decorator class to the hostDirectives array. If you always want to use the input's default value (given the input has a default value), you don't have to declare the input in the hostDirectives array:

```
@Component({ ........., hostDirectives: [
{directive: TypeDirective, inputs: ['btLibsUiType']},
{directive: SizeDirective, inputs: ['btLibsUiSizeType']}]})
export class ButtonComponent { }
```

After defining the objects in the hostDirectives array, the two directives will be applied automatically when you declare the button or tag component in a template. When using directive composition, you can also alias the input values of the directive:

```
{directive: TypeDirective, inputs: ['btLibsUiType: style']}
```

Now, in your HTML templates, if you want to supply a value for the btLibsUiType input of TypeDirective, you can use the following syntax:

```
<bt-libs-button [style]="'secondary'">XX</bt-libs-button>
```

Using directive composition inside directives works the same as in components. Let's say we have backgroundColorDirective and textColorDirective; we can declare textColorDirective inside the hostDirectives array of backgroundColorDirective. Now, when you use backgroundColorDirective, both directives will be applied, and the inputs of both directives will be exposed, given that you defined the inputs of textColorDirective in the hostDirectives array of backgroundColorDirective.

When using directive composition, you need to use standalone directives. Otherwise, it will not work. Also, each time a component is created, a new instance of all the directives declared in the `hostDirectives` array will be created. Because a new instance of each directive is created for each instance of the host component, you must be careful when using directive composition. When you put too many directives inside commonly used components, your memory usage will blow up and negatively affect the performance of your application.

In this section, you learned about attribute directives, structural directives, directive selectors, and directive composition. We will now move on to the next section of this chapter and start to learn more about transforming values by using Angular pipes.

Transforming values using Angular pipes

In Angular, **pipes** are used to transform values. Angular offers a lot of useful built-in pipes and allows you to create your own. Let's first list the most powerful and commonly used built-in pipes and briefly explain what they are used for:

- `AsyncPipe`: `AsyncPipe` is used to handle asynchronous values in your templates. It automatically subscribes to an **observable** or **promise** and returns the latest value emitted by the observable or promise. `AsyncPipe` unsubscribe automatically because this prevents memory leaks. It's recommended to use `AsyncPipe` as much as possible.

- `UpperCasePipe`: This pipe is used to transform a text value into all uppercase characters.

- `LowerCasePipe`: This pipe is used to transform a text value into all lowercase characters.

- `TitleCasePipe`: This pipe is used to capitalize the first letter of each word.

- `CurrencyPipe`: This pipe is used to transform a number value into a currency value with a currency symbol. You can also control the decimal formatting.

- `DatePipe`: This is used to format date values based on the format you specify.

If you want to explore all the built-in pipes, you can find a complete list at this URL: `https://angular.io/guide/pipes`.

Now that you know about the most commonly used built-in pipes, let's see how you can use pipes.

Using pipes in HTML templates and TypeScript files

Pipes are commonly used in HTML templates, but you can also use them in your TypeScript files. To use pipes in HTML templates, you can use the following syntax:

```
<div>{{currentDate | date}}</div>
```

On the left side, you have a property or value; then, you indicate that you're going to use a pipe with the vertical bar (pipe symbol: |), and on the right side of the pipe symbol, you declare the name of the pipe you want to use; in our example, it is date. If your pipe takes a parameter, you supply the parameter by adding a colon and the value like so:

```
<div>{{currentDate | date: 'YYYY-MM-dd'}}</div>
```

When a pipe takes more than one parameter, you can chain them together by adding another colon and append the value after the colon like so:

```
<div>{{currentDate | date: 'YYYY-MM-dd':'GMT'}}</div>
```

You can chain multiple pipes to a value if you need to apply them. When you chain pipes, they will be executed one by one, from left to right. Chaining pipes is done using the following syntax:

```
<div>{{currentDate | date: 'YYYY-MM-dd' | uppercase}}</div>
```

As mentioned before, you can also use pipes inside your TypeScript files. Although pipes are mostly used inside HTML templates, they can also be useful inside your TypeScript files. You can add pipes to the providers array of your component and then inject the pipes using dependency injection. After injecting the pipe, you can use it in your component class by calling the transform method:

```
const formattedDate = this.datePipe.transform(this.currentDate, 'dd/
MM/yyyy');
```

When using standalone components (as we are doing), you need to import the pipe into the component before you can use the pipe. You can either import CommonModule, which contains all the pipes, or import the specific pipe if it's a simple component and you don't need CommonModule for other purposes.

Now that you know how to use pipes in your HTML templates and TypeScript files, let's learn about pure and impure pipes.

Is it pure or impure?

Angular pipes come in two flavors: **pure** and **impure** pipes. It's important to understand the difference between the two because it can affect your application's performance and expected behavior. By default, pipes are pure unless you add the pure false flag. The difference between pure and impure pipes lies in their update behavior and how Angular runs change detection on them.

You need to use a pure transform function when creating a pure pipe. A **pure function** is a function that always returns the same output when given the same input.

Angular only runs a pure pipe when a pure change in the input value is detected. Pure changes are changes to a primitive value (number, string, Boolean, bigint, symbol, undefined, and null) or when a new reference object is supplied (date, array, function, or object). Changes to a reference object are not seen as pure changes. So, if you have a pure pipe that takes an array as a value, updating the array will not trigger the pipe because this is an impure change. When you assign the property using a new array, the pipe will run because it receives a new reference object instead, which is a pure change.

Angular skips updates of reference objects when running pure pipes because detecting pure changes is much faster than performing deep checks on objects; because of this, Angular can quickly determine if your pipes need to be executed again or if the pipe can be skipped. If Angular had to do a deep check or run your pipes on each change detection cycle, it would hugely impact the performance of your application.

So, remember that when you use pure pipes with reference objects, you might not always get what you expect unless you know what you're doing. For example, suppose that you have a dashboard array and a pipe that filters the array only to include active dashboards like this:

```
<div *ngFor="let dashboard of dashboards | active">
    {{dashboard.name}}
</div>
```

Now, when updating the dashboard array with `push`, the pipe will not run because the reference of the dashboard array did not change. If you assign a dashboard's property using a new array, the reference changes and Angular's change detection will trigger the active pipe and filter the results as expected.

When using an impure pipe, Angular will execute the pipe each time it detects a change. This means Angular will run the pipe upon each keystroke or mouse movement. Impure pipes can be useful and will update reference objects as expected, but be careful when using impure pipes, as they can dramatically slow down your application. When you use impure pipes, you want to set the change detection strategy of your component to `OnPush` so that your pipes will not be executed too often. When the component change detection is set to `OnPush`, change detection will only run when the component receives new input values or when you trigger it manually. It's good practice to set your change detection strategy to `OnPush` as much as possible, as it will help improve your application's performance.

Now that you understand the difference between pure and impure pipes, let's learn more about `AsyncPipe`, as it is the most important built-in pipe Angular provides us with. After learning about `AsyncPipe`, we will learn how to create custom pipes.

Using AsyncPipe

The most powerful built-in pipe is `AsyncPipe`. Even though `AsyncPipe` is an impure pipe, it's recommended to use it as much as possible to handle the observable and promise results used in your templates. Using `AsyncPipe` offers advantages over handling observables with subscriptions in your component class.

First, `AsyncPipe` subscribes and, more importantly, unsubscribes to observables automatically. This is very important because it prevents memory leaks. If you don't clean up a subscription correctly, you will end up with memory leaks, and your application will start to slow down and show unexpected behavior up to the point of crashing.

To demonstrate `AsyncPipe`, we will create an observable in the component class using the RxJS `interval` operator like so:

```
timer: Observable<number> = interval(2000);
```

This `interval` observable will emit the next index every 2 seconds, starting at 0. So, after 2 seconds, the observable emits 0, and after another 2 seconds, the observable emits 1, and so on.

We can subscribe to the `interval` observable inside the component class and assign the result to a component property that we display inside the template:

```
this.timer.subscribe((n) => { this.count = n; });
```

Next, you can use the `count` property inside your template:

```
<div>{{count}}</div>
```

If you're using `OnPush` change detection for your component, you need to call a change detection manually each time the observable receives a new value. Otherwise, the `count` property will not be updated in your template. When using the aforementioned approach, you must also add logic to unsubscribe from your observable when the component is destroyed or when the observable property is assigned to another observable. Now, let's see how we can use the interval observable in our HTML template using `AsyncPipe`:

```
<div>{{timer | async}}</div>
```

As you can see, using `AsyncPipe` is simple. You declare the property assigned using the observable (in our case, it's named `timer`) and add `AsyncPipe` next to it. Each time the interval observable emits a new value, it will be reflected in our template. There is no need for an extra property to save the observable result, no need to unsubscribe, and there is no risk of memory leaks! Even when you assign the `timer` property using a new observable, the async pipe will automatically unsubscribe from the old observable and subscribe to the new observable.

When using the async pipe, using `OnPush` change detection is recommended because `AsyncPipe` is impure. Another advantage of `AsyncPipe` is that it automatically marks the component template that needs to be checked for changes when the pipe receives a new value. This is useful when you set your change detection strategy to `OnPush`. When using a regular observable subscription, the HTML template is not marketed to be checked for changes if you use the `OnPush` strategy, meaning you have to trigger the change detection manually after your subscription receives a new value.

Now you know more about `AsyncPipe` and why it's such a powerful tool, let's explore how you can create your own pipes.

Building your own pipes

When creating pipes in our Nx monorepo, we will do so in a `util` library. For our example, we will create a simple pipe that will multiply a number using a specified factor. The correct place to create this pipe is in a library under the shared domain and the `util` type. Use our custom generator to create a new library with the name: `common-pipes` and select **shared** as its domain and **util** as its type.

When your new library is generated, follow these steps to generate the custom pipe:

1. Close and reopen VSCode so that your new library is included in the Nx schematics.
2. Right-click on the folder at this location: `libs\shared\util\common-pipes\src\lib`, and select **Nx generate**.
3. Type `pipe` and click on **@nx/angular – pipe**.
4. Enter `multiply` for the **name*** field.
5. Click on **Show all options**.
6. Check the **standalone** checkbox.
7. Click on **Generate** in the top right.
8. After the component is generated, add the following to the `index.ts` file in the library:

    ```
    export * from './lib/multiply.pipe';
    ```

After that, you can use your pipe, but before using it, let's add some logic.

When you open the `multiply.pipe.ts` file, you'll see that Nx generated a `MultiplyPipe` class, which implements the `PipeTransform` interface. The class is also decorated with the `@Pipe()` decorator, where the standalone flag is set to `true` and the pipe name `multiply` is defined. Inside this decorator, you can add the pure flag; your pipe is pure by default. You only have to add the pure flag with a `false` value when you want to create an impure pipe. In our example, we will create a simple, pure pipe, so there is no need to add the pure flag to the decorator.

Nx also added a transform function to the class to adhere to the `PipeTransform` interface. The transform function is the "heart" of your pipe, where you add your transformation logic. You can adjust the transform function to this:

```
transform(value: number, multiplier = 2): number {
  return value * multiplier;
}
```

As you can see, we have a value and a multiplier function parameter. The value parameter is what we declare on the left side of the pipe in our HTML templates. `multiplier` is the parameter we supply after the colon. We gave `multiplier` a default value of 2, so it's optional when declaring the pipe in your templates.

When you want to use the pipe, you first need to import it; if you work with NgModules, the pipe needs to be imported in an NgModules instance; if you work with standalone components, like we are doing here, you have to import the pipe into the components you want to use it in. After you import the pipe into a standalone component, you can use it in the template like this:

```
<div>{{10 | multiply}}</div>
```

If you want to supply the pipe with a custom multiplier value, you can use the following syntax:

```
<div>{{10 | multiply: 5}}</div>
```

If you want to add more parameters to your pipe, you can do so by adding more parameters inside your transform function. Let's say you want another multiplier in your `multiply` pipe. You can add it as follows:

```
transform(value: number, multiplier = 2, additional = 1): number {
  return value * multiplier * additional;
}
```

Now, you can use the following syntax inside your HTML templates:

```
<div>{{10 | multiply: 5: 10}}</div>
```

Our example pipe is simple, but you can add any logic you want inside your transform function. Just make sure you use a pure function when creating a pure pipe, that is, a function that returns the same value when given the same input and doesn't affect any other code. When you create an impure pipe, make sure not to add time-consuming or resource-intensive code, as it will impact the performance of your application negatively. Here is an example of how you make a pipe impure:

```
@Pipe({
  name: 'multiply',
  standalone: true,
  pure: false,
})
```

Now you know that pipes are used to transform values. Angular offers built-in pipes for common transformations and to handle asynchronous values. You know the difference between pure and impure pipes, and you can create your own custom pipes. To finish this chapter, we will learn about Angular animations.

Creating and reusing stunning animations

In the previous sections, you've seen how to manipulate DOM elements using directives and how to transform template values using pipes; in this section, you will learn how to create animations for your HTML elements and components using the built-in animation module.

To start, you have to enable the animation module. To do this, go to `app.config.ts` in your applications under the *expenses-registration application* in your Nx monorepo. Inside `app.config.ts`, you'll find the `appConfig` object used in your `main.ts` inside the `bootstrapApplication` function. To enable the animation module, add the `provideAnimations()` function inside the providers array of the `appConfig` object like so:

```
provideAnimations(),
```

If you're using an NgModule-based application, you need to import `BrowserAnimationsModule` inside the NgModule where you want to use animations. After adding either the `provideAnimations` function or `BrowserAnimationsModule`, you can start adding animations inside your components. To demonstrate animations, let's create a selectable label component inside our `common-components` library. Use the Nx generator to create a component, name it `selectable-label`, choose the `common-components` library for the project, check the **standalone** checkbox, and set **changeDetection** to OnPush. When the component is generated, add the following export in the `index.ts` of the `common-components` library:

```
export * from './lib/selectable-label/selectable-label.component';
```

Now, add the following code to the `component` class:

```
@Input() labelText!: string;
@Input() get selected() {
  return this._selected;
}
set selected(selected) {
  this._selected = selected;
  this.animationState = selected ? <selected> : <deselected>;
}

@Output() selectedChange = new EventEmitter<boolean>();
private _selected = false;
animationState = 'deselected';

onSelectionChanged() {
  this.selected = !this.selected;
  this.selectedChange.emit(this.selected);
}
```

Add the following CSS to the SCSS file:

```
span {
  color: white; background-color: #455b66;
  padding: 5px 15px; border-radius: 15px; cursor: pointer;
}
```

Add the following HTML to the HTML file:

```
<span (click)="onSelectionChanged()">{{labelText}}</span>
```

Now that we have a simple label component, let's create our animation. Start by adding the `animations` array inside your component decorator like this:

```
@Component({ ........., animations: [] })
```

Animations for the component are added inside this `animations` array. Add the following animation inside the array:

```
trigger('selectedState', [
  state('selected', style({ backgroundColor: '#382632' })),
  state('deselected', style({ backgroundColor: '#455b66'})),
  transition('selected <=> deselected', [animate('2s')])
])
```

This is a simple animation that changes the background color from the hex color #382632 to #455b66 and takes 2 seconds to perform the transition. Now let's examine what we added line by line.

Animation trigger

Our animation starts with a trigger:

```
trigger('selectedState', [])
```

The trigger receives two arguments: the trigger name and an animation metadata array. The trigger name is used to identify the animation and apply it to the HTML elements in the template. The name can be anything you like, but it is recommended to make it descriptive in terms of what your animation does. In our example, we used the name `selectedState`. The animation metadata array contains state and transition functions that define the behavior of our animation. Let's explore these functions in more detail.

Animation state

Inside our animation metadata array, by using our `selectedState` trigger, you'll find the state functions:

```
state('selected', style({ backgroundColor: '#382632' })),
state('deselected', style({ backgroundColor: '#455b66'})),
```

Your animation can have as many state functions as are needed. You can define animations without a state or with many state functions. Each state defines a state your animation can transition to. If you have an animation without any state functions, you can define the style changes inside the transition functions, and the animation will still run, but after it finishes, the HTML element will be as it was before the animation started. When you have a state, you can transition an element from one state to another. When the animation transitions to a specific animation state, the HTML element will stay styled as the state defined it until the animation transitions the HTML element to another animation state.

Each animation state receives a name to indicate the state and a style function to define the style properties with which to transition to for the specific animation state. It is important to note that the styles are indicated with camel case, so no hyphens are used. The background color CSS property becomes `backgroundColor`.

In our example, we have two states: `selected` and `deselected`. Inside the component class, we also have a property called `animationState`, which holds the current state of our animation. By default, it's set to `deselected`. When we click on our label, we will set the `selected` property, and inside the setter of our `selected` property, we will set `animationState` to its proper value.

Animation transition

After our state functions, we define a transition function inside the animation metadata array:

```
transition('selected <=> deselected', [animate('2s')])
```

Transition functions specify how to transition from one animation state to another and they can take three parameters: the transition statement, an animation metadata array, and an object that can define a delay for your transition.

Transition expression

The `transition` expression indicates what state transition to cover when using a specific `transition` function. The syntax of the `transition` expression reads from left to right and uses arrows to indicate the state transition. For example, `selected => deselected` would target state transitions from `selected` to `deselected`, `deselected => selected` would target state transitions from `deselected` to `selected`, and `selected <=> deselected` would target state transitions from `selected` to `deselected` and from `deselected` to `selected`. You can also use an asterisk inside your selection expression. The asterisk symbol is a wildcard and stands for every state.

For example, `* => deselected` would trigger the transition if any state transfers to `deselected`, and `* => *` would trigger if any state transfers to another.

The `transition` expression has a few more special selectors that are similar to the asterisk. For example, you can use `:enter` and `:leave` as transition expressions. The `:enter` expression will be applied to an element entering the DOM, and `:leave` targets elements that are removed from the DOM. The `:enter` and `:leave` expressions do not care about the animation state an HTML element currently has. These two expressions are useful when combined with the `*ngIf` or `*ngFor` directives.

You can also use `:increment` and `:decrement` as expressions. The `:increment` and `:decrement` expressions will trigger the animation when the value inside the HTML element is a number, and it gets incremented or decremented.

Animation metadata array

After the transition expression, the `transition` function also takes an animation metadata array as input. In our example, we only declared an `animate` function inside, which indicates how long the transition takes, which is 2 seconds in our case. The animate function can also take a `style` function like this:

```
animate('2s', style({ color: 'red' })),
```

The `style` function inside the `animate` function is useful if you have no states for your animation or if you want to perform additional animations during the transition that will not last once the state transition is finished.

You can also define keyframes for your `animate` function. With keyframes, you can indicate different stages of the animation; the offset defines how far into the animation you are, with 0 defining the start and 1 defining the end of the animation:

```
animate('2s', keyframes([
    style({ backgroundColor: <blue>, offset: 0}),
    style({ backgroundColor: <red>, offset: 0.8}),
    style({ backgroundColor: <#754600>, offset: 1.0})])),
```

Besides the `animate` function, the animation metadata array inside the transition function can take more configurations. The most commonly used are `group` and `sequence`.

The `sequence` function is used to trigger multiple animate steps one after another. These can be steps that lead to the result of the state you are transitioning to or just additional animation steps that are not included in your animation state:

```
sequence([
    animate(<2s>, style({ backgroundColor: <#382632> })),
    animate(<2s>, style({ color: <orange> }))
])
```

The `group` function is used to group different `animate` functions. When you group `animate` functions, they will be executed simultaneously during the transition. Each group is executed one after another:

```
group([
    animate(<2s>, style({ color: <white> })),
    animate(<2s>, style({ backgroundColor: <#455b66> })),
])
group([
    animate(<2s>, style({ fontSize: <24px> })),
    animate(<2s>, style({ opacity: <0.5> })),
])
```

Now that you know how to define animations inside your component class, let's examine how you can add those animations to HTML elements inside your template.

Adding animations to your template

Adding animations inside your HTML template is pretty straightforward. To add our animation to the selectable label, we need to add the following code line on the span tag:

```
[@selectedState]="animationState"
```

On the left-hand side, you define the animation trigger, which is preceded by an @ sign and enclosed by square brackets. On the right-hand side, you declare the animation state; in our case, we used the `animationState` property from our component class for this, but you can also add ternary operations.

You can also trigger events when the animation starts or finishes by adding the following code to your HTML tag with the animation defined on it:

```
(@selectedState.start)="onAnimationEvent($event)"
(@selectedState.done)="onAnimationEvent($event)"
```

Lastly, you can disable an animation within HTML child elements based on a Boolean value and the `@.disabled` animation control binding like this:

```
<div [@.disabled]="isDisabled"> ......... </div>
```

In the preceding example, all animations inside div with @.disabled are disabled if the isDisabled property is true.

Now that you know how to create and use animations, let's explore how you can reuse animations.

Reusing animations

Creating animations, especially complex ones, can be a lot of work. If you want to apply them within multiple components, you don't want to add duplicated code and create the same animation multiple times. To reuse animations, you can create an animations.ts file in your application or a utils library, depending on your use case. You can create exported functions in this file. Here is an example of our animation as a reusable animation:

```
export function selectedAnimation(): AnimationTriggerMetadata {
  return trigger('selectedState', [
  state(<selected>, style({ backgroundColor: <#382632> })),
  state(<deselected>, style({ backgroundColor: <#455b66>})),
  transition(<selected <=> deselected>,[animate(<2s>)]),])
}
```

Now, inside the component class, you can define the animation as follows:

```
@Component({
  .........
  animations: [selectedAnimation()],
})
```

This maintains readability in your component and allows you to use the animation inside multiple components without creating it again.

Summary

In this chapter, you learned the difference between structural and attribute directives. You learned how to add and remove DOM elements using directives. You've learned to change the styling and behavior of DOM elements using directives, and you now know how to listen out for the events of host elements. You can use built-in directives and create your own. Besides directives, you learned how to transform values using Angular pipes. You learned about pure and impure pipes and how they can be made and impact your performance. Lastly, you made your own Angular animation and learned about animation triggers, states, and transform functions. Now, you know how to declare animations in your templates and how to reuse them throughout your application.

In the next chapter, you will learn about reactive and template-driven forms.

4

Building Forms Like a Pro

Forms are at the core of many frontend applications. They allow you to collect user input to save and act upon data provided by your application users. Building forms is one of the aspects where Angular excels. Angular provides two different approaches for building forms, both with tools to validate individual form fields as well as the validity of the entire form.

In this chapter, you will learn how to create Angular **template-driven** and **reactive forms**, synchronize your form fields with your application state, validate forms and individual form fields, and learn about advanced concepts such as dynamic form creation. By the end of this chapter, you will be able to create forms like a pro!

This chapter will cover the following main topics:

- Understanding the different types of forms in Angular
- Building template-driven forms
- Building reactive forms
- Creating forms dynamically

Understanding the different types of forms in Angular

Angular provides two different types of forms: template-driven forms and reactive forms. This section will discuss the differences between template-driven and reactive forms and help you assess what approach best fits your situation.

Let's start by explaining the key characteristics of both approaches for building forms.

Characteristics of template-driven forms

Template-driven forms are a way to build forms where the HTML template plays a central role. Template-driven forms implicitly define form controls and validation rules in the HTML template using directives. This means the form controls and validation rules are created and managed by the directives you place on the HTML elements instead of manually creating the form controls and validations inside your TypeScript files and directly communicating with the form API yourself.

The template-driven approach relies on two-way data binding using the `ngModel` directive to synchronize changes in the data model for both user input and programmatic changes. When you add the `ngModel` directive to an HTML element, the directive creates and manages a `FormControl` instance for you. These `FormControl` classes that are created by the `ngModel` directive are used to track and validate the status of individual form fields inside your form.

When using template-driven forms, the form data is mutable, meaning you don't update the form values using the form API, but you directly change values that are used within the `FormControl` instances.

> **Important note**
>
> It's also important to note that the data flow for template-driven forms is asynchronous and updates through events and subscriptions. While this asynchronous behavior is managed for you, it's good to know because it can affect tests and sometimes lead to unexpected behavior if you aren't aware that the data flow is asynchronous.

Template-driven forms are easy and quick to build and work very well combined with Signals (we will learn more about Signals in *Chapter 7*), which can reduce the number of times change detection is triggered, improving performance.

However, the data model and form controls of template-driven forms are complex to reuse and test; also, when building large forms, your HTML template becomes bloated. When creating forms that need rigorous testing, reusable form models and controls, or forms that need a dynamic way of being constructed, reactive forms might be the better fit.

Characteristics of reactive forms

Reactive forms provide a flexible, model-driven approach to creating forms. Reactive forms facilitate more programmatic control compared to template-driven forms. When using the reactive approach, you explicitly define the form model by creating `FormGroup`, `FormControl`, and `FormArray` instances in your TypeScript files. The explicit nature of reactive forms allows for more complex logic, easy testing, and better reusability of form controls and models. In reactive forms, validation rules are also explicitly defined using `Validator` classes or custom validators, allowing for more complex validations.

Reactive forms provide fine-grained control over the form's state, allowing you to set, get, and manipulate values programmatically. Once a form and its form controls have been created, you cannot directly modify the values. This makes reactive forms immutable. Immutable forms provide a more reliable data model, which, in turn, leads to fewer bugs.

Because you directly define the form models in your TypeScript files using the form API, the form models that are created for reactive forms can easily be reused and tested. Reactive forms are straightforward when it comes to writing tests because you can directly use the form API in your tests, just like you do in your component classes.

Reactive forms have a synchronous data flow between the view (by *view*, we mean the HTML template that's displayed in the browser) and the data model. As a result, Angular knows precisely when to run change detection on reactive forms, improving performance.

While the reactive approach requires more initial setup compared to template-driven forms, reactive forms are better suited for large, complex forms where additional control, testing, and reusability of form controls and models are crucial.

Key differences between template-driven forms and reactive forms

The following table outlines the differences between template-driven and reactive forms:

	Template-Driven Forms	**Reactive Forms**
Form, model, and validation creation	Implicitly using directives in the HTML template	Explicitly using classes in TypeScript files
Setting up and creating the form	Easy and simple to set up	Needs more initial setup and can feel more complex
Data model	Unstructured and mutable	Structured and immutable
Data flow	Asynchronous	Synchronous
Compatible with Signals	Good	Not good
Testability	Difficult to unit test	Easy to test using the form API
Reusability and dynamic creation of the form	Harder to reuse or build dynamically	Easy to reuse and build dynamically

Table 4.1: Key differences between template-driven forms and reactive forms

Now that you know the key characteristics of Angular template-driven and reactive forms, let's do a deep dive and learn how you can create both types of forms.

Building template-driven forms

In this section, we will build a template-driven form. You will learn how to bind data to input fields, group form fields, and perform built-in and custom validation rules in template-driven forms. You will also learn how template-driven forms work behind the scenes to get a better understanding of template-driven forms.

By the end of this section, you'll be able to build robust template-driven forms and create a template-driven form to add expenses to our demo application.

Creating a forms library with a form component

Before we start creating the form, we need a new library. We will generate the new library using the custom Nx generator we made in *Chapter 1*.

You can debate about how to separate the forms library. You can either create one library that holds all the forms of a specific domain, you can create a forms library for each application in a particular domain, or you can create a new library for each form.

Using a single library for each form is the best way to use the Nx caching and incremental build systems, but it also has some extra overhead in terms of development and maintenance. If your organization has a lot of forms that are reused among multiple applications, splitting them up into individual libraries might be worth the extra setup as it will speed up your builds and pipelines.

For our example, I will create a forms library dedicated to the *expenses-registration application* so that the library will contain all the forms for this specific application:

1. Run the custom Nx generator to create the library. Name it `expenses-registration-forms`.

2. Select **finance** for the domain and **ui** for the type. Then, click **Generate**.

3. Once the library has been generated, restart VS Code so that the Nx schematics are updated with your new library.

4. We will create a component for our template-driven form using the Nx generator. Name the component `add-expense`.

5. Select the newly created library for the project, check the **standalone** checkbox, click **Show all options**, and select **OnPush** for the **changeDetection** option.

6. Click **Generate** in the top-right corner.

7. Once the component has been generated, export it into the `index.ts` file of the library. I like to end the component selector for forms with `form`, so I will rename the component selector `bt-libs-ui-add-expense-form`.

Now that we've created the library and component, we can start creating the template-driven form.

Creating a template-driven form

We will start by creating a simple HTML form and slowly convert it into an Angular template-driven form. Add this to your `add-expense.component.html` file:

```
<form>
  <div class="form-field">
    <label for="description">Description:</label>
    <input type="text" id="description" name="description">
  </div>
  ...........
  <button type="submit">Submit</button>
</form>
```

You can replace the **dots** with any additional form fields you want. In this example, I will have four fields: **description**, **amount excluding VAT**, **VAT percentage**, and **date**.

You can find the complete form and styling in this book's GitHub repository: `https://github.com/PacktPublishing/Effective-Angular/tree/feature/chapter-four/building-forms-like-a-pro`.

Import the `AddExpenseForm` class into `expenses-overview-page.component.ts` and add the component selector in the corresponding HTML file to display our HTML form. When you click the submit button, you'll notice that the page is reloaded; this is the default native behavior when submitting forms. Yet, when building modern applications using frameworks such as Angular, we expect a better user experience where we process form submissions without page reloads, just like when we route to pages without reloading the page.

We'll start converting our native HTML form into an Angular template-driven form by importing `FormsModule`. This `FormsModule` contains all the directives we need to build our template-driven forms. If you're building with `NgModules`, you must import `FormsModule` in the corresponding NgModule. We're using standalone components, so we'll import `FormsModule` into components where we build a form. Add `FormsModule` inside the `imports` array of the component decorator of our newly created `add-expense` component:

```
imports: [CommonModule, FormsModule]
```

After adding `FormsModule`, the page won't be reloaded when you click your form's submit button, which is precisely what we wanted.

But how can this be, since we didn't change our form? The Angular `ngForm` directive prevents the default browser behavior when submitting the form, and because of that, the page isn't reloading. But we didn't add the directive to our form, so why isn't the page reloading?

The answer lies in the selector of the ngForm directive. When we inspect the selector of the ngForm directive, we can see that the directive is automatically applied to all form tags, and because we've added FormsModule, the ngForm directive is applied to the form in our component. Even though the directive is automatically applied to HTML form tags, it's advised to write declarative code and manually apply the directive to form tags. So, add the directive to your form tag like this:

```
<form #addExpenseForm="ngForm">.........</form>
```

On the left, we added a template variable called #addExpenseForm, while on the right, we assigned this template variable with an instance of the ngForm directive. By assigning the directive to a template variable, we can use the directive in all sibling and child elements of our HTML form tag.

Now that we've added the forms module and the ngForm directive, we can start to configure the fields of our template-driven form.

Configuring template-driven form fields

To have a template-driven form, we need to connect the form's input fields to an object in our component class and bind the fields to our ngForm instance to validate the entire form.

Let's start by creating an interface for the object we want to use in our form.

In the add-expense folder, create a new file, add-expense.interface.ts, and add the interface reflecting the fields in your form.

For our example, this is the interface definition:

```
export interface AddExpense {
    description: string;
    amountExclVat: number | null;
    vatPercentage: number | null;
    date: Date | null;
}
```

Also, export the interface in the index.ts file of the library. Now, in add-expense.component.ts, define a property of the AddExpense type, as follows:

```
@Input() expenseToAdd: AddExpense = { description: '', amountExclVat:
null, vatPercentage: null, date: null }
```

Here, we used an input with a default value. We used an input because the form will be a dumb component, and the parent component will input any default values other than empty values.

Note that you don't have to use an object in a template-driven form; you can also use separated properties or a combination of properties and objects. Once you've defined the object, you need to bind the properties of the expenseToAdd object to the fields of your form. We can do this using the ngModel directive.

When you add the ngModel directive, behind the scenes, Angular registers FormControl in the ngForm instance. The ngModel directive allows for two-way data binding, meaning the values of the expenseToAdd object and our form will automatically be synchronized if we change the properties in our component class or if the user enters values inside the form inputs.

You connect the properties of the expenseToAdd object to the form inputs by adding the ngModel directive and the name attribute to each of the form inputs, like this:

```
<input [(ngModel)]="expenseToAdd.description" type="text"
id="description" name="description">
```

As you can see, we use the banana-in-the-box syntax for the ngModel directive, just like we did with two-way data binding for inputs and outputs in *Chapter 2*.

Just like using two-way data binding for input values, you can also split this ngModel into a separate input and output, like this:

```
[ngModel]="expenseToAdd.description" (ngModelChange)="expenseToAdd.
description = $event"
```

> **Important note**
>
> Using the input and output separately can be useful when you're working with Signals or when you want to perform additional logic before binding the input value to the data model.

When you create your input fields, adding the name attribute to the field is important as this provides a unique key for the ngForm instance to track the form field. You will get an error in your browser console if you don't add the name attribute. Now, go ahead and add the ngModel directive and name attribute for all the fields in your form. After adding all the ngModel directives and name attributes, you can confirm that two-way data binding works by temporarily adding the following code to your HTML template:

```
{{ expenseToAdd | json }}
```

After adding the preceding code to your HTML template, the expenseToAdd object is displayed as JSON on your screen. When you start to type in the form inputs, you can see the properties of the expenseToAdd object being updated. Vice versa, when you assign values to the properties of expenseToAdd in your component class, the form is also updated, and just like that, you've created a form with two-way data binding.

The ngModel and ngForm directives have some other interesting configurations that are useful. Let's start by examining ngModelOptions.

Additional form field options

You can use `ngModelOptions` to configure form control instances in template-driven forms. The `ngModelOptions` directive can be used to define the `name` attribute, control the update behavior, or mark an `ngModel` instance as standalone.

You can add `ngModelOptions` by adding the directive to an input field where you declare the `ngModel` directive:

```
[ngModelOptions]="{name: 'description', updateOn: 'blur', standalone:
false}"
```

Let's learn more about the properties you can set on the `ngModelOptions` directive.

Using the name property

When you set the `name` property of `ngModelOptions`, you can remove the `name` attribute on the `input` field as using the `name` property with `ngModelOptions` is the same as providing the `name` attribute.

Using the updateOn property

Next, we have the `updateOn` property, which controls the update behavior of the form control and can take three values – `change`, `blur`, or `submit`:

- `change`: This is the default value and doesn't have to be set explicitly unless the update strategy for the entire form is changed to something else. With the `change` value, the form control will update on each keystroke.

- `blur`: When you set `updateOn` to `blur`, the form control will update when you blur (focus out) from the correlating input field. This can be useful when you run some logic that can take a lot of time or resources to run. If you run heavy logic on each keystroke, your performance will suffer.

- `submit`: If you set `updateOn` to `submit`, the form control will only be updated when the form is submitted. Just like `blur`, the `submit` option can help improve the performance of your form.

You can also change the update behavior for all the form controls in your form by using the `ngFormOptions` directive on the HTML form tag. The `ngFormOptions` directive only takes the `updateOn` property. When you change the update behavior of the entire form, you can still overwrite it for individual form controls using the `ngModelOptions` directive.

Using the standalone property

Lastly, we have the `standalone` property. When you mark a form control as standalone, the form control won't register itself in the `ngForm` instance. When a form control is marked as standalone, you don't have to provide the `name` attribute because the `ngForm` instance doesn't have to track the value of the input. Determining the form's validity will also not account for the fields marketed as standalone. This can be useful when you have form fields that do not represent the form model.

Another scenario for standalone form controls is a single input for which you don't need to create an entire form, but you do want to use the ngModel directive. Some examples might be a theme toggle, language selection, or search.

Grouping template-driven form fields

Often, you have a group of form fields that belong together. A good example is address fields such as street, ZIP code, and house number. You might want to check the validity of the fields as a group, add styling or user feedback, or perform validation logic on the group instead of the individual fields.

For these use cases, Angular provides the `FormGroup` class. In template-driven forms, you can create a `FormGroup` class using the `ngModelGroup` directive. When you declare the `ngModel` group, Angular will create a `FormGroup` class behind the scenes.

Declaring `ngModelGroup` is simple! Just add the `ngModelGroup` directive to an HTML tag, which wraps multiple HTML elements with an `ngModel` directive declared on it.

As an example, we can group our amount, excluding VAT, and VAT percentage fields.

We can do this like so:

1. Wrap the HTML for the amount and VAT percentage fields inside a `fieldset` tag and declare the ngModelGroup directive on this `fieldset` tag.

2. Assign ngModelGroup with a name for your `FormGroup`, and that's it:

   ```
   <fieldset ngModelGroup="amount"></fieldset>
   ```

Now that all the input fields have been declared and bound to your data fields using the `ngModel` directive, you need a way to listen for the form submission.

Submitting template-driven forms

As you may have noticed, we added the `submit` type to our submit button:

```
<button type="submit">Submit</button>
```

This will trigger a submit event. When this submit event happens, Angular will trigger its own internal submit event called ngSubmit. We can listen for this ngSubmit event as we would for any other browser event or component output:

```
<form #addExpenseForm="ngForm" (ngSubmit)="onSubmit()">
```

You declare ngSubmit on the form tag; on the right-hand side, you declare the ngSubmit event in round brackets; and on the left-hand side, you define a function that's declared in the component class to handle the submission.

For our example, we will create an output event to output a clone of the form object and reset the form using the ngForm instance. To achieve this, we need a way to access the ngForm instance in our component class; we can do this using the @ViewChild directive:

```
@ViewChild('addExpenseForm') form!: NgForm;
```

Next, we need to add the component's output, which will allow us to send the submitted values to the parent component so that it can act upon it accordingly:

```
@Output() addExpense = new EventEmitter<AddExpense>();
```

Lastly, we need to add the onSubmit method with our logic:

```
onSubmit() {
  this.addExpense.emit(structuredClone(this.expenseToAdd));
  this.form.reset();
}
```

As you can see, we used the native structuredClone method to send a clone of the expenseToAdd object to the parent. We sent a clone because expenseToAdd is a reference object, meaning it will be cleared everywhere (also in the parent component) after we call reset on the form, and it will clear the object we bound to the ngForm instance.

Now that we've added the submit logic, we'll start exploring validation rules, control status, and how to display messages to the user based on the control statuses and validity of the form and its controls.

Using built-in validation rules for template-driven forms

For a good user experience, it's important to provide good user feedback on our form. You want to indicate when your form fields are valid or still require some changes to be made by the user. You also want to prevent the user from submitting the form when not all the fields are valid and provide good error messages for incorrect or incomplete form fields.

To achieve these feats, you must add form validations to the form fields and use the control statuses to add or remove styling and error messages. Let's start by exploring how to validate form fields in template-driven forms.

Validating the form fields of template-driven forms can be done using directives. Most validations can be done using the built-in validation directives, but when needed, you can also create your own directives to validate template-driven form fields.

You can find all built-in validation directives on the official Angular website: `https://angular.io/api/forms/Validators`.

Let's learn how to add validators to our example form. We will start by making our form fields required using the `required` directive. You can apply the `required` validator by adding this directive to your `input` fields:

```
<input required [(ngModel)]="expenseToAdd.amountExclVat" >
```

As you can see, adding the built-in validators is simple. Let's add another validator to limit the maximum number we accept on our VAT percentage field. It wouldn't make sense to enter a number higher than 100 for a percentage field. To achieve this, you can use the built-in `max` validator. You can still type a number higher than 100, but the form field will be invalid.

To apply the `max` validator, add the following to the input tag:

```
[max]="100"
```

You can also conditionally add validators. Upon using the notation with square brackets, as we did with the `max` validator, and supplying a property to it as a value when the value is set to `null`, the validation rule is disabled:

```
[max]="null" [disabled]="null"
```

If you want, you can add additional validation rules to your form; go through the validators in the Angular documentation and add the directives of the validation rules you want to use.

I briefly want to mention the pattern validator as this is a special one. The pattern validator can be used for many use cases as it takes a regular expression and checks if the value in the input field matches the regular expression pattern. Other validators are used for a single purpose, such as checking for a maximum input value or whether the field has a value.

Now that you know how to add the built-in validation rules, let's find out how we can style the form fields based on the status and validity of the form and form controls.

Styling the form and form fields based on control status values

To provide a good user experience for your application users, it's important to provide visual feedback about the status of the form and its fields. The best way to do this is by utilizing the control status of your form and its form controls.

The form, as well as its `FormGroup` and `FormControl` instances, are updated by Angular with a control status and corresponding CSS classes. The best way to provide visual feedback about the status of your form and its fields is by providing styles for the control status CSS classes. Because these stylings are shared among all your applications, it's a good practice to create them as a global styling.

Let's start by creating a new `global-styling` folder under the shared library folder. In this `global-styling` folder, create a file named `form-control-status.scss`. After you've made the folder and file, add the following under `devDependencies` in the root `package.json` file of the *Nx monorepo*:

```
"@global/styling": "file:libs/shared/global-styling"
```

Next, you need to run `npm install` so that you can use `@global/styling` to import your global styling inside CSS files throughout your *Nx monorepo*.

Now, it's time to add the styling for the control status CSS classes inside our newly created CSS file.

Angular provides us with eight different control status CSS classes by which we can style our form and its form groups and controls:

- `ng-valid`: This is applied when the form control or group is valid based on the validation rules. It's applied to the form when all groups and controls are valid.

- `ng-invalid`: This is applied when the form control doesn't pass all validation rules. It's applied to the form when one or more controls are invalid.

- `ng-pending`: This is applied when an asynchronous validation is being validated.

- `ng-pristine`: This is applied to a form control that has not been interacted with. It's applied to the form when there has been no interaction with any of the form controls.

- `ng-dirty`: This is applied to a form control that has been interacted with. It's applied to the form when there has been interaction with at least one form control.

- `ng-untouched`: This is applied to a form control that the user has not focused on or interacted with. It's applied to the form when no field has been focused or interacted with.

- `ng-touched`: This is applied to a form control that the user has focused on or interacted with. It's applied to the form when at least one field has been focused or interacted with.

- `ng-submitted`: This is applied to the form element when the form has been submitted.

You can use these CSS classes to style the form however you see fit. We will style the form fields based on their validity in our `add-expense` form but only do so when the form has been touched.

You can add the following inside the `form-control-status.scss` file:

```scss
form.ng-touched {
  .ng-valid:not(fieldset)   {
    border-left: 5px solid #42A948; /* green */
  }
  .ng-invalid:not(form, fieldset) {
    border-left: 5px solid #a94442; /* red */
  }
}
```

Now, we only need to import this file into the CSS file of our `add-expense` form. Open your `add-expense.component.scss` file and add the following `import` statement at the top of the file to import the `form-control-status` styling:

```scss
@import '@global/styling/form-control-status';
```

After adding the import, the styles will be applied to the form.

Now that you know how to style the form and its form groups and controls using the control status classes, let's learn how to display messages to the user based on the status of the form controls.

Displaying messages based on the status of form controls

You can improve the user experience even further by showing messages when a form field isn't valid. You can do this by using the form control instances created by `ngModel`.

Start by adding a template variable to the inputs where you declared `ngModel` and assign the `ngModel` instance to the template variable so that you can use it throughout the template:

```html
<input required [(ngModel)]="expenseToAdd.description"
       #description="ngModel" type="text" id="description">
```

As you can see, we've added the `#description` template variable and assigned it with `ngModel`. Now, you can access and control the form control instance created by the `ngModel` instance through the `#decscription` template variable.

Let's add a `span` element with an error message underneath the input element:

```html
<span>This field is required</span>
```

Next, we want to ensure this message is only shown when the form has been touched, and the input validator needs to be satisfied. We can do this by using `ngForm` and the form control instance, which we bound to the template variables. The form control instance exposes several properties that we can check to determine the status of the correlating input field.

Like the control status CSS classes, the form and form control itself expose properties to assess the same statuses: `valid`, `invalid`, `pending`, `pristine`, `dirty`, `touched`, and `untouched`. The form control also exposes `disabled` and `enabled` as checks of the disabled status of the correlating input field.

Besides these Boolean properties, the form control exposes multiple values and methods; you can find all the properties on the official Angular website: `https://angular.io/api/forms/FormControl`.

Now, without further ado, let's add a `*ngIf` statement to our span element so that it's only displayed when the input form is touched and the required validator of the form control isn't satisfied:

```
*ngIf="addExpenseForm.touched && description.hasError('required')"
```

As you can see, we use the `hasError` method to check for the required validator. You can go ahead and add template variables for `ngModel` and error messages for the other fields in the form. For our VAT percentage field, we will add an additional message to check for the `max` validator:

```
<span *ngIf="addExpenseForm.touched && vatPercentage.
hasError('max')">The max percentage allowed is 100</span>
```

In this instance, we use `hasError` with the `max` parameter to get the validity of the `max` validator. If you want to display the message under other conditions, you can change the `*ngIf` statement however you see fit.

Another common way to improve the user experience of your form is to prevent the user from submitting the form when the form isn't valid yet. You can do this by binding the form validity to the `disabled` attribute of the submit button:

```
<button [disabled]="addExpenseForm.invalid" type="submit">Submit</
button>
```

Here, we simply used the reference to the `ngForm` instance and made the button invalid when the `invalid` property of the form returned `true`.

Now that we've enhanced the user experience of our form by adding validation rules and supplying visual user feedback through styling and error messages, let's move on and see how we can create custom validators for template-driven forms.

Custom validators for template-driven forms

To finish this section, we will learn how to create custom validators for template-driven forms. The built-in validators will cover most scenarios, but in some cases, you need to create custom validators – for example, when you want to check values in your database or state when you need to perform cross-field validations or need to perform other validations that can't be done with built-in validators.

All the built-in validators for template-driven forms are directives, so we also need to create a directive to build custom validators.

Let's start by creating a new library with our custom Nx generator. Name the library `form-validators`; then, select **shared** for the domain and **util** for its type. When the library is created, generate a directive inside it. Name the directive `max-word-count` and check the **standalone** checkbox. After creating the directive, create a `template-driven-validators` folder and move the directive inside it. Now, add an export for the directive inside `index.ts`:

```
export * from './lib/template-driven-validators/max-word-count.
directive';
```

Now, let's add some validation logic inside our directive. We'll start by creating an input for our custom validator directive:

```
@Input('btLibsUtilMaxWordCount') maxWords = 1;
```

Next, we must implement the `Validator` interface:

```
export class MaxWordCountDirective implements Validator {}
```

The `Validator` interface requires you to implement a `validate` function:

```
validate(control: AbstractControl): ValidationErrors|null{}
```

As you can see, the `validate` function takes `AbstractControl` as input and returns either `null` or a `ValidationErrors` object. Inside the `validate` function, we will add our validation logic. When the validation passes, we will return `null`, and when the validation doesn't pass, we will return a `ValidationErrors` object. To perform some validation logic, we need to access the form control of the HTML element on which our directive is declared.

The `AbstractControl` function parameter gives you access to a `FormControl` or `FormGroup` instance, depending on which HTML element you place the directive. When you place the directive on HTML elements that declare ngModel, you will receive a `FormControl` instance; if you put the directive on an HTML element with ngModelGroup declared, you will receive a `FormGroup` instance.

In our example, we will use the directive on elements with ngModel on it and receive a `FormControl` instance through the `function` parameter. We want to check the value of our form control and determine if more words are in the value than we defined in our maxWords input. When there are more words, our validation fails, and we return a `ValidationErrors` object; otherwise, we return `null`, which means our validation passes.

First, we must check the number of words:

```
const wordCount = control?.value?.trim().split(' ').length;
```

As you can see, we use `control.value` to get the value from the form control. If there is a value, we trim it and split it into spaces to get the number of words in the string. Next, we check if the number of words is bigger than the `maxWords` input; if that is the case, we return the error object; otherwise, we return `null`. The `error` object can have any format you like:

```
return wordCount > this.maxWords ?    {btLibsUtilMaxWordCount: { count:
wordCount }} : null;
```

That is all the logic we need for our custom validator. But there is still one problem: when we declare the directive on our form, the form won't know this directive is a custom validator; it will think it's just a regular directive. Instead, we want our form to register the directive as a validator so that the form takes the directive into account when determining the form's validity and its form groups and controls. We can achieve this by adding a provider to the directive decorator:

```
providers: [{
   provide: NG_VALIDATORS,
   useExisting: MaxWordCountDirective,
   multi: true
}]
```

By adding the `NG_VALIDATORS` provider to our directive, the `ngForm` instance will register the directive as a validator and include the directive when determining if the form and its groups and controls are valid.

To use the directive, you need to add the directive class' name to the imports of our `add-expense` component since both the directive and our component are standalone. After adding the directive to the imports array, you can use the directive in the HTML template:

```
<input [btLibsUtilMaxWordCount]="3" [(ngModel)]="expenseToAdd.
description" ..........>
```

Now, when you type more than three words in the description input, the form field becomes invalid. You can check if the form control has errors related to our custom directive by using the template variable concerning the description form control and calling the `hasError` method on the control, just like we did when we checked for the required error:

```
description.hasError('btLibsUtilMaxWordCount')
```

When you look at the `error` object we return in our directive, you might notice that we included a `count` property with the number of words we used in our form control:

```
{btLibsUtilMaxWordCount: { count: wordCount }}
```

You can retrieve the `count` value by using the `getError` method on the form control instance:

```
description.getError('btLibsUtilMaxWordCount').count
```

As you can see, we used the description template variable, which holds the reference to the form control instance, and called `getError` to retrieve the error object. You can use the `error` object to display extra information to the user in an error message; in our example, you can include the current count of words.

With that, you know how to add a custom validator to template-driven forms and show error messages when your custom errors occur. Next, we will learn how to validate form groups with custom validators.

Validating form groups with custom validators

In our current example, we're declaring our custom validator directive on HTML elements with `ngModel` declared on it. However, in some cases, you might want to perform cross-field validations or validate multiple fields at once. A typical example is when you have a password and confirm the password input field in your form and want to check if both fields hold the same value.

When you want to perform validation logic on a group of fields, you group the fields in your form using `ngModelGroup`, as we did with our VAT fields. Next, you must declare the custom validator on the HTML tag with `ngModelGroup`.

When you declare the directive on an element with `ngModelGroup` on it, the validator directive will receive `FormGroup` as the `AbstractControl` function parameter. Inside the custom validator, you can then access the values of the individual form controls of the form group, as follows:

```
const password = control.get('password').value;
const confirm = control.get('password-confirm').value;
```

The `string` value inside the `get` method should be equal to the name you declared on the form control in your HTML template:

```
<input required [(ngModel)]="formObj.password" name="password">
```

For the rest, the custom validator is equal to our example validator. You perform the validation logic you want and return `null` if the validation passes and a `ValidationErrors` object when the validation fails.

Async validations with custom validators

Lastly, you can create asynchronous validators. These asynchronous validators work similarly to regular custom validators; there are only two differences. First, the provider for the asynchronous validators is different. If you want to create an asynchronous validator, you must change the provider to the following values:

```
providers: [{
  provide: NG_ASYNC_VALIDATORS,
  useExisting: UsernameAvailabilityDirective,
```

```
    multi: true,
  }]
```

Secondly, you need to return `Promise` or `Observable` with either a `null` value or the `ValidationErrors` object. Here's an example of such a function:

```
validate(control: AbstractControl): Promise<ValidationErrors | null> |
Observable<ValidationErrors | null> {
  const username = control.value;
  return checkUsernameInDatabase(username).pipe(
    map((isAvailable) => (isAvailable ? null : { usernameTaken: true
}))
  );
}
```

In this example, the `checkUsernameInDatabase` function is an API call that returns an observable, and we use RxJS's pipe and map to map the result to `null` or the `ValidationErrors` object. You can use any asynchronous logic inside your validator, so long as you return `Promise` or `Observable` with `null` or the `ValidationErrors` object.

In this section, you learned how to create template-driven forms. You also learned how to create form controls using the `ngModel` directive, how to create form groups using `ngModelGroup`, and how to use built-in validators. Then, you learned about how to style a form and display error messages based on form control statuses. Lastly, you learned how to create custom validators for form controls, form groups, and asynchronous validation rules. Next, we will start learning how to build reactive forms.

Building reactive forms

In this section, you will learn how to build reactive forms. We will rebuild the form we used in the previous section but reactively. You will learn how to create `FormGroup`, `FormArray`, and `FormControl` instances, how to validate reactive forms, and how to create custom validators for reactive forms. You will also learn how to dynamically create form fields and how to change the update behavior of reactive form fields.

Creating a reactive form

Start by removing or commenting out the HTML template and TypeScript code for our template-driven form. I will comment out the code so that it remains an example of the template-driven approach.

Next, we'll start with the same simple HTML form, including the description, amount excluding VAT, VAT percentage, and date fields. We will gradually transform the simple HTML form into a reactive form:

```
<form>
  <div class="form-field">
    <label for="description">Description:</label>
    <input type="text" id="description">
  </div>
  ............
  <button type="submit">Submit</button>
</form>
```

After creating the simple HTML form, we will start by importing ReactiveFormsModule inside our component. When building reactive forms, you import ReactiveFormsModule instead of the regular FormsModule. After adding ReactiveFormsModule to your component file, we can move on and start to create the form model using the FormGroup and FormControl classes.

At its core, an Angular form is a FormGroup class with FormControl elements inside FormGroup. When we created our template-driven form, Angular created a FormGroup class for our ngForm instance and added a FormControl element inside the FormGroup class for each ngModel directive we declared.

To construct our reactive form model, we need to do the same only manually:

```
addExpenseForm = new FormGroup({
  description: new FormControl(''),
  amountExclVat: new FormControl(null),
  vatPercentage: new FormControl(null),
  date: new FormControl(''),
});
```

As you can see, we've created a FormGroup instance and added a FormControl element inside for each form field. Inside the function brackets of the form control, we've added either an empty string or null; these are the default values for the form control instances.

If you want different default values, you can change the values inside the function brackets of the FormControl instances. Later in this section, we will create an input by which you can send default values from the parent, just like we did with the template-driven form.

After creating the form model, you need to bind the form model to the form inside your HTML template. You can bind the form model to the template by using the FormGroup directive and assigning the directive with the form model, as follows:

```
<form [formGroup]="addExpenseForm"> ......... </form>
```

Next, you need to bind the `FormControl` instances to the form field by using the `formControlName` directive:

```
<input formControlName="description" type="text" …… >
```

You must assign the `formControlName` directive with the key that was used to assign `FormControl` inside the `FormGroup` class you declared inside your TypeScript file. After adding the `formControlName` directive to all your form fields, the form model is bound to the HTML form.

Now, when you change the values inside the input fields, `addExpenseForm` in your component class will be updated, and when you update the `FormControl` values inside your TypeScript file, the changes you made will be reflected inside the input fields in the browser. You can test this by adding a default value to one of the `FormControl` values inside your TypeScript file; when you do, the value should be reflected inside the HTML template.

To test if changing the input also changes the values of `addExpenseForm` in your component class, you can temporarily add the following code to your template so that you can see the changes to the `addExpenseForm` object in real time:

```
<div>{{addExpenseForm.value | json}}</div>
```

After creating the reactive form, you can remove the aforementioned line of code. Until then, it can be helpful to see if all values are synchronized.

Now that we've defined and tested our form fields, let's learn how we can group form fields into reactive forms.

Grouping fields in reactive forms

In our template-driven form, we grouped the amount excluding VAT and VAT percentage fields using the `ngModelGroup` directive. By grouping fields, you can perform validation logic on the group instead of the individual fields, style the group, or just change the data structure to something better resembling your state or DTO objects.

Let's also group the amount excluding VAT and the VAT percentage fields in our reactive form. Start by changing the HTML template and wrap the two input fields inside a fieldset HTML template:

```
<fieldset> ……… </fieldset>
```

Next, change the form model so that it reflects the structure where the two fields are grouped inside a FormGroup class:

```
addExpenseForm = new FormGroup({
  description: new FormControl('Test'),
  amount: new FormGroup({
    amountExclVat: new FormControl(null),
```

```
    vatPercentage: new FormControl(null),
  }),
  date: new FormControl(''),
});
```

When you've updated the form model, you need to bind the form group inside the HTML form. You can bind the form group using the `formGroupName` directive and assign it with the key of your `FormGroup` class inside the form model. In our case, this is `amount`:

```
<fieldset formGroupName="amount"> ......... </fieldset>
```

That is all you need to do. Now, when you log the form value, its structure will look like this:

```
{
  "description": "",
  amount": {
    "amountExclVat": null,
    "vatPercentage": null },
  "date": ""
}
```

Now that you know how to group fields using the `FormGroup` class, let's explore how we can dynamically add fields and `FormControl` instances to our form and form model.

Dynamically adding fields and FormControl instances

Unlike template-driven forms, when creating reactive forms, you can also group fields using the `FormArray` class. The `FormArray` class is useful when you don't know how many values will be supplied, and the user can dynamically add and remove input fields in the form. An example of this might be tags or comments.

To demonstrate this, we will add another field to our reactive form so that we can add tags. Start by adding the `FormArray` class to your form model:

```
tags: new FormArray([ new FormControl('')])
```

As you can see, `FormArray` takes an array instead of an object as a parameter. Inside this array, we declared a form control, which will be the first tag inside our form. Because `FormArray` takes an array, we can dynamically add (or remove) from controls, which, in turn, will add or remove input fields to/from our HTML form.

Now that we've added the `FormArray` class inside our form model, let's add some HTML so that the user can add multiple tags.

Start by creating a `fieldset` value and declare the `formArrayName` directive on the HTML tag:

```
<fieldset formArrayName="tags"> </fieldset>
```

Inside `fieldset`, add the following HTML:

```
<div class="form-field" *ngFor="let item of addExpenseForm.controls.
tags.controls; index as i">
  <label for="tag-{{i}}">Tag:</label>
  <div class="tag-field">
   <input [formControlName]="i" type="text" id="tag-{{i}}">
  </div>
</div>
```

The aforementioned HTML will output a label and input for each `FormControl` inside `FormArray`. Now, we only need a way to add and remove `FormControl` instances.

Let's add two buttons to the HTML underneath the input tag:

```
<div>
  <button *ngIf="i > 0" (click)="removeTag(i)">-</button>
  <button (click)="addTag()">+</button>
</div>
```

Next, we need to add the logic for the click functions inside the component class:

```
addTag() {
  this.addExpenseForm.controls.tags.insert(0, new FormControl(''));
}
removeTag(index: number) {
  this.addExpenseForm.controls.tags.removeAt(index);
}
```

With the aforementioned functions added to the component class, you can now add and remove `FormControl` instances inside the `FormArray` class, which will result in added or removed input fields.

But what if you have data in an object format instead of an array and you need to keep the form format, and just like our previous example, you don't know how many or what fields you will have and what the keys for these fields will be?

An example could be when you receive a list of statuses that our expense has gone through, such as submitted, waiting for revision, checked, approved, and so on. For these scenarios, you can use the `FormRecord` class. We won't add this to our form, but I will outline an example of how to handle a `FormRecord` class:

```
statuses: new FormRecord({})
```

The FormRecord class receives an object with keys and FormControl or FormGroup instances. You can add the FormControl instances to the FormRecord class as follows:

```
this.form.controls.statuses.addControl('someKey', new
FormControl(''));
```

As you can see, we utilize the addControl method, which is exposed on the FormRecord class. The addControl method is also available on FormGroup, but you should use FormRecord when you don't know what fields will be added beforehand. You can use the FormGroup when you do know what the entire model will look like. You can strongly type FormGroup and FormRecord. To strongly type FormGroup, use the following syntax:

```
address: new FormGroup<IAddress>({..........})
```

In this example, we added an interface for FormGroup between the arrow brackets. By adding the type, you strongly typed FormGroup and can't add fields not defined in the IAddress interface. When strongly typing a FormRecord class, we tell the record what value our dynamic controls will have. For example, if we have a list of keys with a Boolean value, we can use this syntax:

```
statuses: new FormRecord<FormControl<boolean>>({..........})
```

Here, we tell FormRecord that each field that will be added should be a FormContro instance and that FormControl should have a Boolean value. If we have form controls with different values, we can change the Boolean value for string or any other type our controls will hold.

Now that you know how to group fields using FormGroup or dynamically add fields using FormArray, FormGroup, or FormRecord, let's explore how we can control the update behavior of our fields and how to declare standalone controls in a reactive form.

Configuring update behavior and declaring standalone controls

With template-driven forms, we have the option to configure the update behavior of form control instances using the ngModelOptions directive; in reactive forms, we control the update behavior inside the FormGroup, FormArray, or FormControl class. You can add a configuration object to set the updateOn property. You can set the updateOn property for FormGroup, FormArray, and FormControl elements alike.

When you set the updateOn property for a FormGroup element of FormArray, it will be applied to all nested FormControl elements. If you define the updateOn property for a nested object inside the form model, that nested property will overrule the updateOn property of parent elements. Just like template-driven forms, you can set the update behavior to change (which is the default), blur, or submit:

```
description: new FormControl('', {updateOn: 'blur'}),
```

When you used the `ngModelOptions` directive inside our template-driven form, you also had the option to mark an input field as standalone. As with template-driven forms, you can also have a standalone reactive form element, but you don't have to set it with a standalone property; instead, you just declare a `FormControl` instance without `FormGroup`:

```
searchInput = new FormControl('');
```

For a standalone reactive form field, you can use the `formControl` directive instead of the `formControlName` directive inside your HTML template:

```
<input [formControl]="searchInput" type="text">
```

You can access the value of your standalone form field by using the value property of `FormControl`:

```
this.searchInput.value;
```

Instead of using the value property, you can also access the value more reactively and react to each update of `FormControl`. You can handle the changes reactively by subscribing to the `valueChanges` observable:

```
this.searchInput.valueChanges.subscribe(() => { ......... });
```

Now that you know how to control the update behavior of the form fields and how you can reactively create standalone form fields, let's learn how to programmatically set and update form values.

Setting and updating values programmatically

Often, you need a way to set and update values programmatically inside your component class. In reactive forms, you can use `setValue` to set values on individual form controls and the `patchValue` method when you want to update multiple fields of your form simultaneously.

In this section, we will create our component `@Input()` directive and use `patchValue` to update the default form values with the values we received from the parent component.

Start by adding a new interface inside the `add-expense.interface.ts` file:

```
export interface AddExpenseReactive {
  description?: string;
  amount?: {
    amountExclVat?: number;
    vatPercentage?: number;
  };
  date?: string[];
  tags?: string[];
}
```

Next, we will add the `@Input()` directive with a setter. Inside this setter, we will use the `patchValue` method:

```
@Input()
public set expenseToAdd(value: AddExpenseReactive) {
  this.addExpenseForm.patchValue(value);
}
```

The `patchValue` method will update all values that are supplied inside the value object. So, if the value only contains the description key, only the description will be updated; when the value object contains the description and amount with both properties, all these values will be updated. The only exceptions are the `date` and `tags` fields.

As you may have noticed, when we defined the date in the interface, we gave the `date` property a string array type; this is because to set a default value, we need to supply the form control with a string array, like this:

```
['2023-10-15']
```

If you provide a value in a similar format from the parent component, `patchValue` will also work for the date; when you provide it with a simple string, the input will not be populated.

Make sure you also update the default value inside `FormControl`; otherwise, you will get compiler errors because the types inside your control and patch value don't match:

```
date: new FormControl(['']),
```

Besides the `date` field, the `tags` field is also different because we use it to add controls to our form dynamically. When we assign `addExpenseForm` with the form model, our `FormArray` tags receive a default value of one `FormControl`. Because we only added one `FormControl` inside `FormArray`, when we use the `patchValue` method on the form, only one tag will be set, even if more tags are supplied. To update the value of `FormArray`, we need to add some additional logic inside our setter:

```
this.addExpenseForm.controls.tags.clear();
value.tags?.forEach(tag => {
  this.addExpenseForm.controls.tags.push(new FormControl(tag)); });
```

First, we used the `clear` method on the `FormArray` tags. The `clear` method will clear all existing `FormControl` instances declared inside the `FormArray`. After we clear the `FormArray`, we will use a `forEach` loop to add a new `FormControl` for each of the tags we received from the parent component.

Now, when we supply an object with values from the parent component, our form will be populated with these values.

In some scenarios, you only want to set the value of a single control. When you only want to set the value of a single control, you must use the `setValue` method on the `FormControl` instance:

```
this.addExpenseForm.controls.description.setValue('New description');
```

The `setValue` method doesn't allow you to assign a number value to the description input field, making it a type-safe and programmatic way to set the values of form controls. We added the input to receive values from the parent component.

With that, you've learned how to set and update values programmatically for your reactive forms. Next, we will start learning about validation in reactive forms.

Validating reactive forms

As with template-driven forms, you can validate reactive forms using built-in or custom validators. We'll start by looking into the built-in validators and then create a custom validator. Reactive forms have the same built-in validators as template-driven forms, except we don't declare them using directives but the `Validator` class. Inside your `FormControl` instances, you can add an array with the validators you want to apply.

To add the required validator to our description field, we can use this syntax:

```
description: new FormControl('', [Validators.required]),
```

That's all we need to do to add the built-in validators – simple and clean. When we want to create custom validators for our reactive forms, we must create a function that returns the `ValidatorFn` interface. We will create the same max word count validator, just like we did with the template-driven form, now only as a function instead of a directive.

To create the custom validator, start by creating a new folder, `reactive-validators`, inside the `form-validators` library. Inside this folder, create a new file named `max-word-count.function.ts`. We will use this new file to create our validator function:

```
export function maxWordCountValidator(maxWords: number): ValidatorFn {
  return (control: AbstractControl): ValidationErrors | null => {
    const wordCount = control?.value?.trim().split(' ').length;
    return wordCount > maxWords ? { maxWordCount: { count: wordCount }
} : null; }; }
```

Here, we create the `maxWordCountValidator` function, which will receive an input for the maximum word count we will allow. Inside this function, we return an implementation of the `ValidatorFn` interface. Here, `ValidatorFn` is the same as the `validate` method we declared inside the directive; we check the word count and return `null` if the word count is equal to or smaller than the allowed count and return a `ValidationErrors` object otherwise.

Next, you can add this custom validator to your form controls:

```
description: new FormControl('', [Validators.required,
maxWordCountValidator(3)]),
```

Now, when you type four or more words inside the description input, the form field becomes invalid. With reactive forms, you can also dynamically add and remove validators. Compared to template-driven forms, dynamically adding validators is simpler because you don't have to think about different null values for different validators; you simply call the addValidators or removeValidators method:

```
this.addExpenseForm.controls.description.addValidators(Validators.
required);
this.addExpenseForm.controls.description.removeValidators(Validators.
required);
```

With that, you've learned how to use the built-in validators, how to create validators using the ValidatorFn implementation, and how to add and remove validators dynamically. In the next section, you'll learn how to provide visual feedback in reactive forms.

Providing visual feedback about the form's state in reactive forms

Angular applies control status CSS classes to form elements, just like it does with template-driven forms. The control status CSS classes are the same ones that are used for template-driven and reactive forms, so we don't have to change anything for the styles we've already created to be applied. The FormGroup, FormArray, FormRecord, and FormControl instances will all receive the control value CSS classes based on their current status.

The only real difference is how we display the error messages. We created a template variable in the template-driven form and bound it to the ngModel instance to access the form controls. When we are using reactive forms, we access the form control instances through the form model we created:

```
<span *ngIf="addExpenseForm.touched && addExpenseForm.controls.
description.hasError('required')">This field is required</span>
```

For the rest, nothing changes, so go ahead and add the error messages you want to display to your users. We already discussed providing visual feedback when we covered template-driven forms, so when it comes to reactive forms, this is all we have to cover.

Submitting and resetting a reactive form works the same as for template-driven forms, so you can copy the submission and reset behavior from the *Building template-driven forms* section. To finish this chapter, we will learn how to build forms based on a configuration object dynamically.

Creating forms dynamically

Creating a good form requires quite a bit of code. Regardless of whether you're using the template-driven or reactive approach, you need a lot of HTML; you need to define the model, add validators, and additional logic such as the submit behavior.

You could use a base class for some of the shared functionality, but you can also build a dynamic form, which will dynamically build the form based on a JSON input. In this section, we will build a simple example of a dynamic form. You can extend the dynamic form to fit your specific needs. For example, you might want to fetch the configuration from an external source or support additional validators.

To start our dynamic form, create a new form component named `dynamic-form` in your `expenses-registration-forms` library. Next, create a `dynamic-control.interfaces.ts` file. I will create the new interface inside the component folder, but you can locate all interfaces in a designated folder or use any other folder structure you like. Our new dynamic control interfaces will define the interface for a form control, which will be generated dynamically:

```
export interface DynamicControl {
  controlKey: string;
  formFieldType?: 'input' | 'select';
  inputType?: string;
  label?: string;
  defaultValue?: any;
  selectOptions?: string[];
  updateOn: 'change' | 'blur' | 'submit';
  validators?: ValidatorFn[];
}
```

Once we've defined the interface, we need to add an input property to the component, which will receive an array of `DynamicControl` objects, and a `formModel` property, which will hold our form model:

```
@Input() formModelConfig: DynamicControl[] = [];
formModel = new FormGroup({});
```

When we receive the form configuration as input, we need to build our form model. We can use the ngOnChanges life cycle hook to build the form model each time we receive a new form configuration:

```
ngOnChanges(changes: SimpleChanges) {
  if (changes[<formModelConfig']) {
    this.formModel = new FormGroup({});
    this.formModelConfig.forEach((control) => {
      this.formModel.addControl(
        control.controlKey,
```

```
        new FormControl(control.defaultValue, { updateOn: control.
updateOn, validators: control.validators }));
    }); }}
```

As you can see, we check if the `changes` object contains a new `formModelConfig`; when `formModelConfig` is included in the changes, we use a `forEach` loop to add the form controls of `formModelConfig` to our form model. We also need a submit function and an output that will send the form model to the parent component on submission:

```
@Output() outputForm = new EventEmitter();

onSubmit() {
  this.outputForm.emit(structuredClone(this.formModel.value));
  this.formModel.reset();
}
```

For our component class, this is everything we need. We need to translate this into the template so that our form will be built based on the configuration. Start by adding the form tag, bind it to the form model, and add the `ngSubmit` function to the `form` tag:

```
<div class="form-container">
  <form [formGroup]="formModel" (ngSubmit)="onSubmit()">
  </form>
</div>
```

Next, we need to create an input for each configuration inside `formModelConfig`. We will use a `*ngFor` loop to output elements for each instance inside `formModelConfig` and the `*ngSwitch` directive to determine which element to create. We will use the properties of `DynamicControl` to bind the elements to the form and provide all the correct values:

```
<div class="form-field" *ngFor="let control of formModelConfig">
  <label for="description">{{control.label}}</label>

  <ng-container [ngSwitch]="control.formFieldType">
    <input *ngSwitchCase="'input'"
           formControlName="{{control.controlKey}}"
           type="{{control.inputType}}">

    <select *ngSwitchCase="'select'"
            formControlName="{{control.controlKey}}">
      <option
        *ngFor="let option of control.selectOptions"
        value="{{option}}">
      </option>
```

```
    </select>
  </ng-container>
</div>
```

Once you've added this HMTL, each form element will be rendered and bound to the form model. The last thing we need to do is add error messages. Here's an example of how you can display error messages inside your dynamic form:

```
<span *ngIf="formModel.touched && formModel.get(control.controlKey)?.
hasError('required')"> This field is required</span>
```

Add an additional `span` for each error message your form supports. To test the dynamic form, you can import it inside `expenses-overview-page.component` and add the selector inside the HTML template:

```
<bt-libs-ui-dynamic-form [formModelConfig]="formModelConfig"
(outputForm)="addExpense($event)" />
```

Create `formModelConfig` inside the component class so that the dynamic form has fields to generate:

```
formModelConfig: DynamicControl[] = [
  {
    controlKey: 'description', formFieldType: 'input',
    inputType: 'text', label: 'Description',
    defaultValue: '', updateOn: 'change',
    validators: [Validators.required]
  },
  {
    controlKey: 'amount', formFieldType: 'input',
    inputType: 'number', label: 'Amount excl. VAT',
    defaultValue: null, updateOn: 'change',
    validators: [Validators.required]
  }
]
```

Go ahead and add the rest of the fields we used inside our reactive and template-driven form to `formModelConfig`; you will see that the same form will be generated dynamically, including the validation rules and error messages.

This is just a simple example of a dynamic form; you can add additional logic if you want, such as to allow form groups, form arrays, and form records. The concept stays the same; just adjust the model, add the logic inside the component class to generate the form model correctly, and adjust the template so that you can render it how you intended.

Summary

In this chapter, you learned about Angular's different types of forms. You now know the difference between template-driven and reactive forms and when to use which type. We created a template-driven form that includes validation, error messages, default values, and styling based on control statuses. We also created a custom validator directive for the template-driven form. Next, we recreated the same for using reactive forms.

We also created a custom validator function that can be used inside the reactive form. We learned about dynamically adding fields inside form group, form array, or form record classes inside the reactive form. Then, we learned how to change our fields' update behavior in both template-driven and reactive forms. Lastly, you built a dynamic form that builds the form model based on a configuration and will render the form accordingly, including validations and error messages.

In the next chapter, you will learn how to create dynamic components, which can be reused in many scenarios.

Part 2: Handling Application State and Writing Cleaner, More Scalable Code

In this part, you'll learn how to develop cleaner, more scalable, and performant code for your Angular applications. You'll start by developing dynamic components suited for more complex UI scenarios. You'll learn about lazy-loading individual components on demand to reduce your bundle size and enhance performance. Then, you'll explore commonly used conventions and design patterns to develop more robust and scalable Angular applications. You'll finish this part by getting hands-on experience implementing the facade pattern, state management using NgRx, and reactive programming with RxJs and signals.

This part includes the following chapters:

- *Chapter 5, Creating Dynamic Angular Components*
- *Chapter 6, Applying Code Conventions and Design Patterns in Angular*
- *Chapter 7, Mastering Reactive Programming in Angular*
- *Chapter 8, Handling Application State with Grace*

5

Creating Dynamic Angular Components

When we are creating components, the flexibility and reusability of those components should be top priorities. You don't want unnecessary dependencies inside your component and want to ensure the component can serve as many scenarios as possible without becoming overcomplicated.

This chapter will teach you how to create truly dynamic UI components using content projection, template references, and template outlets. We will learn how to dynamically render components using the component outlet directive and the view container reference.

By the end of this chapter, you will know when and how to project content into UI components, effectively use templates inside your components, and output code in different places, depending on certain conditions. You will also be able to load and render components dynamically, enhancing the flexibility and performance of your application.

This chapter will cover the following main topics:

- A deep dive into Angular content projection
- Using template references and variables
- Rendering components dynamically

A deep dive into Angular content projection

Often, when creating a component, you need to display a wide variety of content inside it. Some good examples include a modal component, a card, or a tab component.

Let's consider the **modal component**; you want the component to have a visible and hidden state, a backdrop, and some shared styling so that all your modals have the same look and feel. Yet the content inside each instance of the modal component will be wildly different. Sometimes, you want to display a form inside the modal, while other times, you want to use it to display text or provide actions or configurations to the user. Most likely, each modal you have within your application will have different content inside.

So, a question arises: how do we facilitate this need and create a component that can house any content it needs within its HTML template?

You could hardcode all the options and use inputs to configure the component, but this becomes unmaintainable really fast! Using **content projection** is the correct way to create a component that needs to display a wide variety of content inside its template. Angular content projection lets you define placeholders inside the HTML template of a component. You can fill these placeholders with any content by projecting the content inside the placeholder from the parent component HTML template where the dynamic component is declared. We will explore and use content projection ourselves by creating a modal component.

Creating a modal component using content projection

Let's start by creating a modal component inside the common-components library of your *Nx monorepo*. Inside the component class, add an input for the shown status of the modal and an output for the close event. Also, add input for the modal title:

```
@Input({ required: true }) title = '';
@Input({ required: true }) shown: boolean;
@Output() shownChange = new EventEmitter<boolean>();
```

Now, let's create the HTML template for the modal component. Start with an ng-container element. Inside the ng-container element sits the modal container and the backdrop. We must also place a *ngIf directive with the shown status on the ng-container element:

```
<ng-container *ngIf="shown">
  <div class="modal-container">
  </div>
  <div class="backdrop"></div>
</ng-container>
```

Next, you can add a modal header and modal content area inside the modal container. The modal header will contain the title and a "X" button to close the modal:

```
<div class="modal-header">
  <h1>{{title}}</h1>
```

```
  <span (click)="shown = false; shownChange.emit()">X
  </span>
</div>
```

Lastly, you have to add the modal content area. This is where we will create the placeholder for our content projection:

```
<div class="modal-content">
  <ng-content></ng-content>
</div>
```

You can find the CSS for the modal component in this book's GitHub repository: https://github.com/PacktPublishing/Effective-Angular/tree/feature/chapter-five/dynamic-components.

As you can see, inside the modal content area, we defined an ng-content element:

```
<ng-content></ng-content>
```

This ng-content element is the placeholder for the projected content. The ng-content element will display everything we project from the parent component into the modal component.

You project content from the parent component into the modal component by placing content between the component's opening and closing selector tags. To test the content projection of your modal, import the modal component inside the expenses-overview component and add a Boolean property, addExpenseShown. Next, add the model component to the HTML template and project addExpenseForm, which we created in the previous chapter, inside the modal component, like this:

```
<bt-libs-modal [(shown)]="addExpenseShown" [title]="'Add expense'">
  <bt-libs-ui-add-expense-form (addExpense)="addExpense($event)" />
</bt-libs-modal>
```

Now, when the modal is shown, it displays the add expense form we projected into the component. The ng-content element inside the HTML template of the modal will be replaced with the add expense form we projected into the modal component.

As you can see, the ng-content slot provides the flexibility you need for your modal component and allows you to project any content easily into the modal. Any logic related to the projected content is handled in the parent component and not inside the modal component itself, resulting in good separations of concerns.

In the case of our projected form, if you want to handle the addExpense output event of the form component, you handle the event inside the expenses-overview component. The same goes for styling the projected content. If you want to style your projected content, you must do so inside the CSS file of the component where you project the content. In our example, this would be the add-expense.component.scss file.

You now know how to project content into a single ng-content slot. Next, you'll learn how to handle more complex projection scenarios by using multiple ng-content slots.

Exploring multi-slot content projection with ng-content select

Using a single ng-content slot provides a lot of flexibility, but sometimes, you need multiple places to project content.

Let's say that in some of the modal designs you have, there is custom content between the title and closing buttons of the modal. To cover this use case, you need two places to project content: one in the content area and one in the header. When using multiple ng-content elements in a component, you need to use ng-content with a select attribute defined on the element:

```
<div class="modal-header">
  <h1>{{title}}</h1>
  <ng-content select="[header-content]"></ng-content>
  <span (click)="shown = false; shownChange.emit()">X</span>
</div>
```

Here, we added the additional ng-content element inside the modal header and assigned the select attribute with a value of header-content. If we added the second ng-content element without defining the select attribute, all projected content would end up inside the last ng-content element in the HTML template of the modal.

For projected content to end up in a specific ng-content element, the projected content needs to match the select attribute value of ng-content. For the header's ng-content element, this means you need to add a header-content attribute to the HTML you want to project to the header slot:

```
<bt-libs-modal [shown]="addExpenseShown" [title]="'Add expenses'">
  <bt-libs-ui-add-expense-form (addExpense)="addExpense($event)" />
  <div header-content>special header content</div>
</bt-libs-modal>
```

You can project as much HTML as you want to a specific ng-content element. There are three scenarios when Angular determines where your content will be projected. Let's explore these scenarios one by one:

- If the projected content matches the select attribute value of an ng-content element in your template, the content is projected to ng-content with the matching select value.

- If the content matches multiple ng-content selectors, the content is projected into the first ng-content element with a matching select value.

- If the projected HTML doesn't match any of the `ng-content` select values, the content is projected to the fallback `ng-content` element – that is, `ng-content` – without a `select` attribute assigned:

```
<ng-content></ng-content>
```

In our modal component, the fallback slot is defined inside the content area of the modal. If all content slots have a select attribute defined, there is no fallback slot. When there is no fallback `ng-content` element, any projected content that doesn't match with at least one of the `select` attribute values will not be projected and rendered.

The `select` attribute of the `ng-content` tag allows you to create powerful selectors that can match various things. You can match HTML attributes like we did in our modal component, but you can also match HTML tags.

For example, let's say we want to project all `div` HTML tags into the header slot. We can adjust the selector like this:

```
<ng-content select="div"></ng-content>
```

Now, you can remove the `header-content` attribute from the projected content, and it will still be projected to the header slot because it's a `div` HTML tag:

```
<div>special header content</div>
```

If you change the `select` attribute value to match a span HTML element, our `div` will end up in the fallback `ng-content` slot inside our content area of the modal component. You can also match on multiple values; for example, if you want to match all `div` and `p` tags, you can create a selector like this:

```
<ng-content select="div, p"></ng-content>
```

You can also match HTML elements on CSS classes, IDs, or specific attribute values. Here is an example of each:

```
<ng-content select=".header"></ng-content>
<ng-content select="#header"></ng-content>
<ng-content select="[type='text']"></ng-content>
```

Using good `select` values for your `ng-content` tags can significantly improve the developer's experience within your team by preventing you and your teammates from looking up the correct selector for the slots each time they have to project content into the component.

Having multiple ng-content slots and making good use of the select attribute on ng-content tags allows for even more flexible components with more control over the projected content. But even with multiple ng-content elements, you might need more flexibility to cover all your component design needs. Sometimes, you need to output your projected content numerous times inside your UI component or conditionally display the content in different places of the HTML template.

Displaying projected content multiple times or conditionally

When you need to display the projected content multiple times or conditionally in the HTML template, your first instinct might be to add ng-content inside a div element and apply the *ngFor, *ngIf, or *ngSwitch directive (or the control flow syntax versions – that is, @for, @if, and @ switch) to the div tag:

```
<div *ngFor="let item of [1,2,3]">
    <ng-content></ng-content>
</div>
```

If you try this, you will find that your content will be projected differently than expected. When using *ngFor, your content will only be projected once, and when using *ngIf or *ngSwitch combined with an ng-content element, only the first rendered ng-content element is displayed. So, the three directives (or control flow syntax) will not work in combination with the ng-content tag.

You can use control flow or the *ngFor, *ngIf, and *ngSwitch directives inside the parent component where you project the content:

```
<bt-libs-modal [shown]="addExpenseShown" [title]="'Add expenses'">
    <div *ngFor="let header of headers" header-content>{{header}}</div>
</bt-libs-modal>
```

Using these directives or control flow in the parent component is good enough for many scenarios, but sometimes, you must use the directives inside the component where the ng-content elements reside. You might need to use the directives inside the component receiving the projected content because of the design needs of the component or to create a better architecture with a good separation of concerns.

For scenarios where you need to use the *ngFor, *ngIf, or *ngSwitch directives or control flow in the component that's receiving the projected content, you don't use ng-content as the projection slot; instead, you need to use the ng-template element.

In this section, you learned how to project content using the ng-content element and where you can run into the limits of what you can do with the projected content when using ng-content. You also learned how to effectively use the select attribute and project content into multiple slots using ng-content.

In the next section, we will learn about ng-template, template variables, and template references.

Using template references and variables

In the previous section, we projected content using the ng-content element. Yet, we ran into a limitation that didn't allow you to use control flow or the *ngFor, *ngIf, or *ngSwitch directives in combination with the ng-content element. We will start by demonstrating how to resolve this using the **ng-template** element. After that, we will learn about other use cases and implementations of the ng-template element and how you can create and use template variables within HTML templates.

We will create a **display-scales** component to demonstrate content projection with an ng-template element and use directives on the project content. The display-scales component is just a simple example to demonstrate the concept of content projection combined with structural directives such as *ngFor.

Start using the *Nx generator* to create the display-scales component next to the modal component. The display-scales component will receive an array of scale sizes as input and display the projected content in these different scale sizes using a *ngFor directive.

When the display-scales component is created, add a scales-projection.directive.ts file to the display-scales folder. Inside the scales-projection.directive.ts file, you can add the following content:

```
@Directive({
  selector: '[btLibsScalesProjection]',
  standalone: true,
})
export class ScalesProjectionDirective {
  constructor(public templateRef: TemplateRef<unknown>) { }
}
```

The scales projection directive only injects the **templateRef** – that's it! The display-scales component uses the directive to access the projected content. Inside the display-scales component class, you need to add the input for receiving the scale sizes and a @ContentChild() decorator to access the projected content:

```
@Input({ required: true }) scaleSizes!: number[];
@ContentChild(ScalesProjectionDirective) content!:
ScalesProjectionDirective;
```

As you can see, we used `ScalesProjectionDirective` as a value for the `@ContentChild` decorator. The `@ContentChild` decorator will get the projected `ng-template` element from the HTML template and hold a reference to the projected content so that we can use the content within the `display-scales` component. Inside the HTML template of the `display-scales` component, you can add this:

```
<div *ngFor="let size of scaleSizes; let i = index" [style.
transform]="'scale(' + size + ')'">
  <ng-container [ngTemplateOutlet]="content.templateRef"></ng-
container>
</div>
```

Here, we created a `div` element that will be rendered for each size inside the `scaleSizes` array the component receives as input. Inside the `div` element, we declared a `ng-container` element with the **ngTemplateOutlet** directive declared on the element.

This `ng-container` with the `ngTemplateOutlet` directive will display the projected content, similar to the `ng-content` element we used in the previous section. The `ngTemplateOutlet` directive needs to receive a `TemplateRef` property; this is why we use the `ng-template` element and the directive that injected `TemplateRef`.

We bind the `TemplateRef` property of the `@ContentChild` decorator value to the `ngTemplateOutlet` directive using `content.templateRef`. When using `ng-template` to project your content, you can use `*ngFor` and other structural directives in combination with the projected content. We couldn't do this when using the `ng-content` element.

Next, you need to project a `TemplateRef` property into the component using `ScalesProjectionDirective`. By doing so, the `@ContentChild` decorator can access the `TemplateRef` property, and we can assign the projected template reference to the `ngTemplateOutlet` directive.

You can project a `TemplateRef` property by using the `ng-template` element. The `ng-template` element is the HTML representation of the `TemplateRef` class. The projected content needs to match the `@ContentChild` decorator, so we need to add the `ScalesProjectionDirective` decorator to the `ng-template` element we are about to project. Don't forget to import the `display-scales` component and `ScalesProjectionDirective` into the standalone component where you are using the scales component and directive:

```
<bt-libs-display-scales [scaleSizes]="[0.8, 1, 1.2, 1.4]">
  <ng-template btLibsScalesProjection>I scale!</ng-template>
</bt-libs-display-scales>
```

After adding the preceding HTML into an HTML template (for example, in the `expenses-overview.component.html` file), you will display the scales component with the `I scale!` content projected.

When you try to project an additional ng-template element, you will notice that only one ng-template element is projected and used inside the display-scales component. Only the first ng-template element is used because we use the @ContentChild decorator inside the display-scales component.

If you want to project multiple ng-template elements with the same projection directive (in our case, ScalesProjectionDirective), you need to use the @ContentChildren decorator instead of the @ContentChild decorator. The @ContentChildren decorator creates a QueryList property of template references instead of a single template reference:

```
@ContentChildren(ScalesProjectionDirective) content!:
QueryList<ScalesProjectionDirective>;
```

To output all elements inside QueryList, you need to loop over QueryList and create an ng-container element for each node in the list. You can achieve this by turning QueryList into an array and using the *ngFor directive to output each item in the list:

```
<div *ngFor="let item of content.toArray()">
  <ng-container [ngTemplateOutlet]="item.templateRef"></ng-container>
</div>
```

If you want to project content into different slots, as we did with the ng-content tags and select attribute, you need to create multiple projection directives like ScalesProjectionDirective and use multiple @ContentChild decorators to separate the project's ng-template elements:

```
@ContentChild(ScalesProjectionDirective) content!:
ScalesProjectionDirective;
@ContentChild(ScalesHeaderProjectionDirective) headerContent!:
ScalesHeaderProjectionDirective;
```

In the HTML template, you can then use the different @ContentChild elements with different ng-container elements, as follows:

```
<ng-container [ngTemplateOutlet]="content.templateRef"></ng-container>
<ng-container [ngTemplateOutlet]="headerContent.templateRef"></ng-
container>
```

With that, you know how you can project ng-template elements and use the *ngFor, *ngIf, and *ngSwitch directives or control flow syntax in combination with projected content.

Now, let's explore what else we can do with the ng-template element, what template variables are, and how we can use them effectively in our components. We will start by learning about template variables.

Using template variables effectively

When building components, we often need to use one part of the HTML template in another part of the HTML template. You can access elements from your HTML template using decorators such as @ViewChild or @ContentChild, but you can also use **template variables**. Using template variables helps simplify your code because you can handle situations where you need one part of the template in another place, all in your HTML; there's no need to create a property in your component class and use this variable inside your HTML template.

A template variable can refer to five different elements:

- DOM elements within the HTML template

- Directives used within the HTML template

- Components used within the HTML template

- TemplateRef from ng-template elements used within the HTML template

- A web component

Template variables are created using the hashtag (#) sign combined with a variable name. For example, if you want to create a template variable from a div element in your template, you can use this syntax:

```
<div #exampleVar> ......... </div>
```

In the preceding example, #exampleVar is the template variable, and the variable holds a reference to the DOM element it's placed on; in this example, the DOM element is the div element. You can now use the exampleVar template variable within your HTML template. When you assign the template variable with a DOM element, you can access all properties of the DOM element, similar to when you access it from within your component class:

```
<div>Template var says: {{exampleVar.innerText}} </div>
```

You can use the template variable with interpolation or anywhere else that you can use component properties inside your template:

```
<input #name placeholder="Enter your name" />
<button (click)="submitName(name.value)">Submit</button>
```

If you want to assign a template variable to a component, you can use the same syntax but only on the selector of a custom component:

```
<bt-libs-modal #modal [(shown)]="addExpenseShown"> .........
</bt-libs-modal>
```

When you use a template variable on a component, you can access all the public properties and methods of the component. Sometimes, it can be useful to access the properties of the child component in this manner, but you should avoid calling public methods like so.

It's not advised to call methods and potentially mutate the data of child components using template variables as it's a bit of an anti-pattern. It would be best if you tried to do all parent-child component communication through inputs and outputs unless there is no other way. Yet it's good to know you can access the properties through template variables so that you can recognize when someone else uses it or when you have a situation where you want or need to use it.

Another common and more accepted use case of template variables is assigning a directive to a template variable. We did this in *Chapter 4* when we created template-driven forms. The syntax for assigning a directive to a template variable differs slightly from assigning a DOM element or component to a template variable. When assigning a directive to a template variable, we must create the variable as normal and assign the directive using the is sign (=):

```
<form #expenseForm="ngForm"> ......... </form>
<div *ngIf="!expenseForm.form.valid">
  <p>Invalid form</p>
</div>
```

In the preceding example, we assigned the ngForm directive to the expenseForm template variable. We can now access the ngForm directive and all its properties through the template variable. Using the template variable makes it easy to use all kinds of form values and statuses such as pristine, dirty, valid, and invalid. There's no need to create a variable within your component class; you can handle everything from within your HTML template.

You can also use template variables to access TemplateRef instances created with ng-template elements. Accessing a TemplateRef instance through a template variable works similarly to assigning a template variable to a component or DOM element; you just add the template variable to the ng-template element:

```
<ng-template #templateOne>I am a template</ng-template>
```

What you can do with a TemplateRef element, either by using the ng-template element (with or without a template variable) or by using the TemplateRef in your component class, is the topic of the next section.

Using TemplateRef elements effectively

Template references can be defined inside your HTML templates using the ng-template element. You can access a TemplateRef element in your component and directive classes using the TemplateRef class. The TemplateRef class and the ng-template element are the same thing in different forms.

A `TemplateRef` element can be used for many use cases. As shown at the start of the *Using template references and variables* section of this chapter, `TemplateRef` can be used for content projection when you need to combine the projected content with a structural directive such as `*ngFor`, `*ngIf`, or `*ngSwitch`.

Another place where `TemplateRef` is used is within custom structural directives. Structural directives add or remove `TemplateRef` elements to/from the view container based on some logic. We added and removed template references to the view container in *Chapter 3* when we created our custom structural directive.

A `TemplateRef` element can also be combined with template context to build truly dynamic components that can display different templates using different data sources. You can also use `TemplateRef` to display content in different places on your page – for example, when you have the same piece of content located on different locations of the page for your mobile and desktop designs.

Another place where the `ng-template` element shines is when you need to render content conditionally based on the outcome of a `*ngIf` directive. Let's continue by exploring the use cases for `TemplateRef` and create some examples.

Combining ng-template with *ngIf

You can display an `ng-template` element based on the outcome of a `*ngIf` statement. Often, you'll see something like this in the HTML templates of Angular applications:

```
<div *ngIf="expenses"> .......... </div>
<div *ngIf="!expenses">Loading...</div>
```

You can do the same thing with the `ng-template` element and define a template that is shown when the `*ngIf` statement isn't met:

```
<div *ngIf="expenses else loading"> .......... </div>
<ng-template #loading>
  <div>Loading...</div>
</ng-template>
```

Instead of declaring the `*ngIf` directive on both `div` elements, we used the `*ngIf else` syntax. We referenced the template variable that was placed on the `ng-template` element for the `else` statement. Now, when there are no expenses, the loading template is shown. This is just a simple example, but if you have large templates to display conditionally, using templates helps separate the HTML. Also, when your `*ngIf` statement uses the `async` pipe, which is often the case, you cannot use the first approach:

```
<div *ngIf="!expensesVm$ | async">Loading...</div>
```

When you use the `async` pipe inside your `*ngIf` directive, you cannot use the logical not operator (!) in your template. You could set a value inside your component class using an RxJS pipe operator on your observable value. Still, it would be a lot simpler and cleaner to use the `*ngIf else` statement and an `ng-template` element, like this:

```
<div *ngIf="expensesVm$ | async as expense; else loading">......</div>
<ng-template #loading>
  <div>Loading...</div>
</ng-template>
```

Using the syntax mentioned previously lets you use the `async` pipe to handle your observable and display an alternative template, so long as no value has been resolved by the observable or the result of the observable has been mapped to a false, null, or undefined value.

You can even take it a step further and split up all your HTML into separate templates using the `*ngIf` `then-else` syntax combined with `ng-template` elements. Using the `if-then-else` syntax gives you a clear separation in your HTML and helps improve maintainability when you have large templates where you have to display multiple blocks of content conditionally:

```
<ng-container *ngIf="expensesVm$ | async; then expenses; else
loading"></ng-container>

<ng-template #expenses> <div>......</div> </ng-template>
<ng-template #loading> <div>Loading...</div> </ng-template>
```

What solution you use is primarily up to your preference and your team's preference. There is no real convention or best practice. When working with the `async` pipe, I like to use the `ng-template` syntax instead of using a pipe operator on my observable to assign a component property, which can be used for the `*ngIf` statement with the logical not operator.

For simple scenarios with synchronous values, I mostly use `*ngIf` and `*ngIf` with the logical not operator to display the correct HTML content – unless there are two large blocks of HTML content, in which case I prefer using templates.

Now that you know how you can use the `ng-template` element combined with the `*ngIf` directive to display templates conditionally, let's explore how to use `ng-template` to display templates in different locations on the same page.

Displaying content in the correct spot using ng-template

Often, you have a design where the same block of content is placed on a different location of the page for your mobile and desktop view:

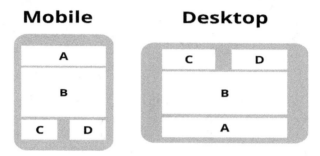

Figure 5.1: Content placement for the mobile and desktop views

As shown in *Figure 5.1*, on mobile, the content blocks, *C* and *D*, are shown below element *B*; on the desktop view, the same content is shown above element *B*. The same goes for block *A*, only in reverse. You don't want to duplicate the HTML for blocks *A*, *C*, and *D* and place it once above and once below block *B*. Duplicating the HTML would break the **Don't Repeat Yourself** (**DRY**) principle and result in a harder-to-maintain and read HTML template.

The ng-template element comes to the rescue in this scenario and lets you display the same template in different places on the page without duplicating the HTML:

```
<ng-container *ngIf="isMobile" [ngTemplateOutlet]="A"></ng-container>
<ng-container *ngIf="!isMobile" [ngTemplateOutlet]="CD"></ng-container>
<div>Block B</div>
<ng-container *ngIf="!isMobile" [ngTemplateOutlet]="A"></ng-container>
<ng-container *ngIf="isMobile" [ngTemplateOutlet]="CD"></ng-container>

<ng-template #A>
  <div> ...... </div>
</ng-template>
<ng-template #CD>
  <div> ...... </div>
</ng-template>
```

As you can see in the preceding HTML snippet, we used one ng-template element to define the HTML for the A block and one ng-template element to define the HTML for the C and D blocks combined. We display the templates using ng-container elements with an ngTemplateOutlet directive referencing the template variables declared on the ng-template elements.

In this example, we used an `isMobile` Boolean to hide or show the correct `ng-container` elements, but you can also wrap them inside a `div` element and hide them with CSS. Using the `ng-template` element combined with the `ng-container` element, we only have to create the HTML for the A, C, and D blocks once, resulting in a cleaner and more maintainable HTML template.

You can clean up the HTML even more and assign the `aboveB` and `belowB` templates inside your component class. I don't like to add the additional logic for this inside my component class, but it's a preference.

Here's an example of how you can handle this in your component class. First, adjust your HTML to this:

```
<ng-container [ngTemplateOutlet]="aboveB"></ng-container>
<div>Block B</div>
<ng-container [ngTemplateOutlet]="belowB"></ng-container>
```

Now, inside your component class, get a reference to the A and CD templates by using the `@ViewChild()` decorator and create a variable for the `aboveB` and `belowB` `TemplateRef` elements:

```
@ViewChild('A') templateA!: TemplateRef<unknown>;
@ViewChild('CD') templateCD!: TemplateRef<unknown>;
aboveB!: TemplateRef<unknown>;
belowB!: TemplateRef<unknown>;
```

Inside the `ngAfterViewInit` life cycle hook, you can assign the variables with the correct template:

```
this.aboveB = this.isMobile ? this.templateA : this.templateCD;
this.belowB = this.isMobile ? this.templateCD : this.templateA;
```

If your component is set to the *OnPush* change detection strategy, you need to manually trigger change detection for the view to update and display the templates. If you don't use the *OnPush* change detection strategy, your view will be updated automatically.

Something similar can be done when you have a page with multiple modals. Instead of defining multiple modal components and projecting the correct content into each modal, you can create one modal that contains an `ng-container` element with the `ngTemplateOutlet` directive and set the correct template when the modal is opened:

```
showModal = false;
modalTitle = '';
modalContent!: TemplateRef<unknown>;

openModal(title: string, content: TemplateRef<unknown>) {
  this.modalTitle = title;
  this.modalContent = content;
  this.showModal = true;
}
```

In the preceding code snippet, we created an `openModel` function, which will be called from the HTML template when one of the modal components needs to be opened. We will pass along a title for the modal and a `TemplateRef` element for the content:

```
<button (click)="openModal('Title A', modalA)">Open modal A</button>
<button (click)="openModal('Title B', modalB)">Open modal B</button>
<bt-libs-modal [(shown)]="showModal" [title]="modalTitle">
  <ng-container [ngTemplateOutlet]="modalContent"></ng-container>
</bt-libs-modal>

<ng-template #modalA>
  <div>A...</div>
</ng-template>
<ng-template #modalB>
  <div>B...</div>
</ng-template>
```

In the HTML, we now only display one modal and display the `TemplateRef` element inside, which is set in the `openModal` method in our component class. When the corresponding button is clicked, we send the `TemplateRef` element from the HTML template to our `openModal` function.

As you may have noticed, the `ng-template` element and `ngTemplateOutlet` can bring a lot of flexibility to your components and let you easily display different templates in different locations. But `ng-template` can offer us even more flexibility by adding context and using `TemplateRef` as an input.

Creating dynamic components using templates and context

We have already seen how we can combine content projection and the `ng-template` element. In this section, we will use `TemplateRef` as a component input and provide additional context to `TemplateRef` to make it even more dynamic. A good example to showcase `ng-template` with context is by creating a dynamic select component, so go ahead and create a select component inside the `common-components` library, next to the `modal` and `display-scales` components.

Once you've created the `select` component, start by adding an input for the select options and a `TemplateRef` element. Also, add an input for the default `selectedIndex` options and an output for when the selection changes. Lastly, add a function to set the selected index when a selection is made and emit the selected value using the output:

```
@Input({ required: true }) options!: unknown[];
@Input() optionTemplate?: TemplateRef<unknown>;
@Input() selectedIndex?: number;
@Input() labelKey?: string;
```

```
@Output() selectedChange = new EventEmitter<unknown>();
onOptionChange(index: any) {
  this.selectedIndex = index.target.value;
  this.selectedChange.emit(this.options[index.target.value]);
}
```

Next, we will add the HTML for our dynamic select component. Start by creating a select element and an option for when there is no selected value. Next, add an ng-container element with a *ngFor directive looping over each option that's received in the options input. We will also track the index of the for loop because we will use the index for the option value. Inside the *ngFor element, we'll define an option element and another ng-container element that defines the ngTemplateOutlet directive to either display a default template or the template received as input. The default template can be used when the select options have no special display needs. In all other scenarios, a template can be provided through the optionTemplate input:

```
<select (change)="onOptionChange($event)">
  <option [value]="null" [selected]="!selectedIndex">Make selection</
option>

  <ng-container *ngFor="let option of options; index as i">
    <option [value]="i" [selected]="i === selectedIndex">
      <ng-container [ngTemplateOutlet]="optionTemplate ||
defaultTemplate"
        [ngTemplateOutletContext]="{ $implicit: option}">
      </ng-container>
    </option>
  </ng-container>
</select>

<ng-template #defaultTemplate let-option>
  {{ labelKey ? option[labelKey] : option }}
</ng-template>
```

As you can see, we also added the ngTemplateOutletContext directive to the ng-container element, which will display the template. The ngTemplateOutletContext directive is assigned to an object containing a $implicit property. The object you assign to the ngTemplateOutletContext directive can be used within the ng-template element, as we did in the preceding code snippet. You can use the $implicit value from the preceding code example inside an ng-template element. To use the $implicit value, you must use the let-propertyName syntax, as shown here:

```
<ng-template let-option></ng-template>
```

You can also assign an object with multiple properties to the ngTemplateOutletContext directive. When you have other context properties besides the $implicit property, you must use a slightly different syntax to use the additional property. Let's say you have a shown property inside the ngTemplateOutletContext directive:

```
[ngTemplateOutletContext]="{ $implicit: option, shown: option.shown}"
```

You must declare the shown property so that it can be used in the ng-template element using this syntax:

```
let-show="shown"
```

The left-hand side can have any name and is the name you use inside the ng-template element; the right-hand side needs to match the property you declared in the ngTemplateOutletContext directive. For example, you can replace let-shown with let-shownValue or anything else you desire, so long as it starts with let-:

```
<ng-template let-shownValue="shown">
  <div>{{shownValue}}</div>
</ng-template>
```

Now that you know how to use the context inside ng-template elements, let's test the select component and see how we can provide a custom template for the select component.

If you're using the select component in another standalone component, you must import the select component before you can declare it in the HTML template. Once you've imported the select component, you can use it in the template like this:

```
<bt-libs-select (selectedChange)="onOptionChange($event)"
[options]="['Test', 'Test 2']">
</bt-libs-select>
```

In a real-world scenario, you should define the input for the options inside the component class, but I've added it directly in the HTML for demonstration purposes. The default ng-template element will be used because we didn't provide a TemplateRef element for the optionTemplate input of the select component.

Now, let's create a custom template to provide to the select component. Let's say you have an array of expenses that looks like this:

```
[{expense: 'Food',amount: 10},{expense: 'Gas',amount: 20}]
```

Now, we can create a template to display the expense and amount inside the `select` component:

```
<ng-template #expenseSelect let-expense>
  <span>Product: {{ expense.expense }}, Amount: {{ expense.amount }}</
span>
</ng-template>
```

When the template is created, you need to assign the template to the `optionTemplate` input of the `select` component, like this:

```
<bt-libs-select (selectedChange)="onOptionChange($event)"
[optionTemplate]="expenseSelect" [options]="expenses">
</bt-libs-select>
```

After providing the custom template to the `optionSelect` input, the select component will use the `expenseSelect` template instead of the default template. The `expenseSelect` template receives the expense object that's used inside the template through the `ngTemplateOutletContext` directive. By using the `ng-template` element and the `ngTemplateOutletContext` directive, you can display any content you want inside the `select` component and use any array of objects that's necessary to provide the data for the template, giving you all the flexibility you need for a truly dynamic component.

We covered a lot in this section. First, you learned about `TemplateRef` and how to use it inside the component class or HTML. Using the `ng-template` and `ng-content` elements, you learned how to use structural directives combined with projected content. You also learned how to provide context to `ng-template` elements using the `ngTemplateOutletContext` directive to build truly dynamic components. Finally, you learned about template variables and how to use them to access values inside your templates or display content conditionally.

In the next and last section of this chapter, you will learn about rendering components dynamically.

Rendering components dynamically

In some scenarios, you might want to load and render components dynamically. You might want to load and render components dynamically because you don't know the layout or exact components of the page upfront or because you have data and resource-intensive components that you only want to load and render if the user needs them in the view.

Here are some common examples of when dynamically loading components is useful:

- When building a website builder where customers can build up web pages based on a set of components you provide. With a website builder, you don't know how the user will create the page's layout and what components will be used. You want to load components dynamically whenever the user adds them to the page.

- When you have a multi-step wizard where the content of the next steps differs based on the users' choices during each step of the wizard.

- Tabs, modals, and popups with resource-intensive components or where different components can be displayed based on the user interaction. You only want to load and render the components if the user requests them.

- When you allow users to configure a list of widgets in your application. You don't know what widgets the user will activate, and in what order they want them displayed, so it would make little sense to load and render them before the user configures them.

- An ad banner component with different ads cycling through the ad banner component. When different teams frequently add new ad components for the banner, a static component structure would make little sense.

When you need to load and render components dynamically, there are three approaches: the ngComponentOutlet directive, the ViewContainerRef class, and, since Angular 17, the defer control flow. First, we'll use ngComponentOutlet as this is the most straightforward solution that works for any Angular version. Next, we will show you how to render and load components dynamically with the defer control flow that was introduced in Angular 17.

Rendering components dynamically using ngComponentOutlet

To demonstrate dynamic component rendering using the ngComponentOutlet directive, we will create a widget component that can render different widgets with their own custom functionality and design. Let's say the widget container can receive widget data as input and render the widget it receives as input; this way, you can display a different widget for each page if needed.

Start by generating a widget-container component inside the common-components library. This widget-container component will dynamically render widget components using the ngComponentOutlet directive.

Next, create two more components: a weather widget and a clock widget component. For demonstration purposes, we will leave the weather and clock widget templates as Nx generated them by default.

Now, in the widget container component, define an ng-container element in the template and add the ngComponentOutlet directive to the ng-container element:

```
<ng-container *ngComponentOutlet="widget.component" />
```

As you can see, we assign the ngComponentOutlet directive with widget.component. So, let's create an interface for the widget and add the widget property to the component class:

```
export interface widget { component: Type<any> | null };
```

Make the widget property an input so that the widget container can receive this widget property from the parent where you declare the widget container component:

```
@Input() widget: widget = {component: null};
```

As a simple example, this is all you need to render a component dynamically. We will extend this example quite a bit, but to showcase the dynamically rendered component, you can now use the widget container by adding the following to the HTML template of one of your components:

```
<bt-libs-widget-container *ngIf="activeWidget"
[widget]="activeWidget"></bt-libs-widget-container>
```

In the component class, you need to add the active widget property that's used for the input of the widget container:

```
activeWidget!: widget;
```

Let's say we want to alternate between the clock and the weather widget every 5 seconds. We can use setInterval for this and assign the activeWidget property with the clock or weather widget:

```
protected readonly cd = inject(ChangeDetectorRef);
showWeather = true;

ngOnInit() {
  setInterval(() => {
    this.activeWidget = { component: this.showWeather ?
WeatherWidgetComponent : ClockWidgetComponent };
    this.showWeather = !this.showWeather;
    this.cd.detectChanges();
  }, 5000)
}
```

When you open your component in the browser, you will see that after 5 seconds, the weather widget is shown, and the widget will alternate with the clock widget every 5 seconds after that. This is, of course, just a simple example and can be improved upon a lot, but it shows how to render the widgets dynamically quite well.

Let's continue and see how we can improve the widget container and add additional flexibility. In some scenarios, your widget components might need data or access to a service to function properly. When the widget needs to do this, you can add an injector to the ngComponentOutlet directive.

Using an injector with ngComponentOutlet

You can provide additional data to your dynamically loaded components using the `injector` property of the `ngComponentOutlet` directive.

Let's start by adding the `injector` property to the `ngComponentOutlet` directive inside the HTML template of the widget container component:

```
<ng-container *ngComponentOutlet="widget.component; injector: widget.
injector;" />
```

Next, you need to update the widget interface so that it contains the `injector` property:

```
export interface widget { component: Type<any> | null; injector:
Injector | any };
```

When the widget container can handle the `injector` property, you must adjust the widget and the input the widget container component receives. First, we will adjust the widget. As an example, we will adjust the weather widget so that it can receive a city and a message that we will display in the HTML template.

Create a new file called `widget-tokens.ts` and add an interface for the weather widget data and an injection token, like this:

```
export interface WeatherWidgetData {city: string; message: string;}
export const WEATHERWIDGET = new
InjectionToken<WeatherWidgetData>('weather widgets');
```

Inside the weather widget component class, you need to inject the `WEATHERWIDGET` injection token:

```
widgetData = inject(WEATHERWIDGET);
```

Now, adjust the HTML template of the weather widget component so that it uses the values of the `WeatherWidgetData` interface:

```
<p>{{widgetData.city}}: {{widgetData.message}}</p>
```

That is everything you need to do inside the weather widget component itself. So, to reiterate, you created an interface and injection tokens, you injected the injection token, and you will receive the injector from the widget container's `ngComponentOutlet` injector property. To close the circle and make everything work, you need to provide the correct input to the widget container and include the injector that will provide the city and message data.

Before you added the injector, you provided the following as input to the widget container to display the weather widget:

```
{ component: WeatherWidgetComponent }
```

To provide the city and message data for the weather widget, you need to add the `injector` property to the input so that the widget container can include it in the `ngComponentOutlet` directive.

You can create the `injector` property by using the `create()` method on the `Injector` class:

```
Injector.create()
```

Inside this `create` method, there's a `providers` object. This is similar to the provider objects you added inside the `providers` array to a component, module, or your application configuration object in your `app.config.ts` file:

```
{ providers: [{ provide: WEATHERWIDGET, useValue: { city: 'Amsterdam',
message: 'Sunny' } }] }
```

For the `provide` property, use the injection token you created inside the weather widget component file. For our example, you'll use the `useValue` property and assign it to the city and message value you want to use. You can create any valid provider object here, so you can also provide services or factory classes instead of the `useValue` property. The entire `Injector.create()` method looks like this:

```
Injector.create({ providers: [{ provide: WEATHERWIDGET, useValue: {
city: 'Amsterdam', message: 'Sunny' } }] })
```

The entire input object for the widget container looks like this:

```
{ component: WeatherWidgetComponent, injector: Injector.create({
providers: [{ provide: WEATHERWIDGET, useValue: { city: 'Amsterdam',
message: 'Sunny' } }] }) }
```

Now, when the weather widget is displayed through the widget container, the `injector` property is passed along, and the city and message values are used within the HTML template of the weather widget. Cool stuff!

Using the `inject` property of the `ngComponentOutlet` directive allows you to provide any service, factory method, or static data to the dynamically rendered components, making that dynamically rendered component as flexible as any other component.

Since Angular version 16.2.0-next.4, you can simplify providing (simple) values to dynamic components a bit by using the inputs property on the `ngComponentOutlet` directive instead of the injector property. To use the inputs property on `ngComponentOutlet`, you need to add `@Input()` properties to the widget. In our example, we can add `city` and `message` `@Input()` properties to the weather widget:

```
@Input() city: string;
@Input() message: string;
```

Once you've added the @Input() properties to the widget, you can provide the @Input() properties with values through the ngComponentOutlet inputs property, like this:

```
<ng-container *ngComponentOutlet="widget.component; inputs:
widgetInputs;" />
```

The inputs property of the ngComponentOutlet directive needs to receive an object; in our case, we named the object widgetInputs. The object has a key for each input property of the component and a corresponding value. So, in our case, the widgetInputs object looks like this:

```
widgetInputs = {
  'city': 'Amsterdam',
  'message': 'Sunny',
}
```

Using the inputs property is a lot easier, but it can only provide objects with simple values. Using the injector property provides more flexibility because it can provide simple values but also classes, services, and everything else you can inject within Angular applications.

Our widget system can already render components dynamically and provide injectors to the dynamically created components. But the dynamically rendered components are still loaded eagerly, so we still need to improve a bit so that we only load the widget components when they need to be rendered.

Lazy loading dynamic components

To improve things further, we can lazy load our dynamic components. Currently, when you load the page, all widgets will be loaded upfront. It would be better if we only loaded the widget components if we needed them, or in other terms lazy loaded the components. We need to change our widget container component so that it lazy loads our dynamically rendered components.

Let's start by creating a new file named widget-loaders.ts. The widget-loaders.ts file will list some types and an object with the import statements for the lazy loaded widgets, a bit like lazy loading components with the router:

```
const widgetKeys = ['weatherWidget', 'clockWidget'] as const;

type WidgetKey = typeof widgetKeys[number];

export type WidgetLoader = { [key in WidgetKey]: () => Promise<any> };

export const widgetLoaders: WidgetLoader = {
  weatherWidget: () => import('../weather-widget/weather-widget.
component'),
  clockWidget: () => import('../clock-widget/clock-widget.component'),
```

```
};
```

```
export type WidgetOption = WidgetLoader[keyof WidgetLoader];
```

We started by creating a `widgetKeys` constant; here, you define all the keys you can use inside the `widgetLoaders` object. Next, we make a type of the `widgetKeys` constant to create a type-safe `WidgetLoader` type. The `WidgetLoader` type defines a key-value pair where the keys can only be values declared inside the `widgetKeys` constant.

Next, we created the `widgetLoaders` object, which types with the `WidgetLoader` type. The `widgetLoaders` object will hold key-value pairs where the keys are values from the `widgetKeys` constant and the values import statements for the lazy loaded widgets.

Lastly, we created a `WidgetOption` type, which allows you to take a single value of the `widgetLoaders` object and nothing else. Now that we've created a type-safe way to define and select widget loaders, we can start to adjust the widget container component.

First, we must remove the widget input we had and replace it with a regular input property. Then, we can add an input for the `injector` property and one for a `WidgetOption` property from the `widgetLoaders` object:

```
@Input() injector!: Injector;
@Input({ required: true }) widgetLoader!: WidgetOption;
widget: widget = { component: null, injector: null };
```

You also need to inject `ChangeDetectorRef` because we have to trigger change detection manually:

```
protected readonly cd = inject(ChangeDetectorRef);
```

Next, you must add the `ngOnChanges` life cycle hook with the `async` keyword in front because we will use async `await` to load the dynamic component:

```
async ngOnChanges(changes: SimpleChanges) {}
```

Inside the `ngOnChanges` life cycle hook, we will get the current value of the `widgetLoader` input:

```
const widgetLoader: WidgetOption = changes['widgetLoader'].
currentValue;
```

Next, we'll use the `widgetLoader` value to lazy load the widget component. When the widget component is loaded, we will assign the widget property using the lazy-loaded component and the `injector` property. Lastly, we need to trigger change detection to reflect the changes within the UI:

```
const widget = await widgetLoader();
this.widget = { component: widget[Object.keys(widget)[0]], injector:
this.injector };
this.cd.detectChanges();
```

With the preceding changes added to the widget container component, everything is in place, and you can lazy load and render widget components dynamically. To test this, we need to use the widget container component inside another component and give it a widget loader and injector as input:

```
<bt-libs-widget-container [widgetLoader]="widget"
[injector]="injector"></bt-libs-widget-container>
```

As an example, if you want to alternate between the clock and weather widget, you can add the following code to the component class where you've added the widget container in the template:

```
widget: WidgetOption = widgetLoaders.weatherWidget;
injector: Injector | null = Injector.create({ providers: [{ provide:
WEATHERWIDGET, useValue: { city: 'Amsterdam', message: 'Sunny' } }]
});
protected readonly cd = inject(ChangeDetectorRef);

ngOnInit() {
  setInterval(() => {
    this.widget = this.widget === widgetLoaders.clockWidget ?
widgetLoaders.weatherWidget : widgetLoaders.clockWidget;
    this.injector = this.widget === widgetLoaders.clockWidget ? null
: Injector.create({ providers: [{ provide: WEATHERWIDGET, useValue: {
city: 'Amsterdam', message: 'Sunny' } }] });
    this.cd.detectChanges();
  }, 5000)
}
```

When you inspect the **Network** tab of your browser's developer tools, you will see the weather and clock widget components being loaded when they enter the view for the first time. When the widgets are displayed again, and the components are already loaded, the browser won't load them again because the browser caches them for you. Now, you have a truly dynamic widget system where you can lazy load and render widgets on demand.

Rendering components dynamically using the defer control flow

In Angular 17, the **defer control flow** was added, which allows you to load and render components easily based on specific triggers. Using the defer control flow can be useful in many scenarios, such as rendering components when they enter the viewport or dynamically rendering components based on conditions. To showcase the defer control flow, we will create a new component named `defer-widget`.

This `DeferWidgetComponent` will have the same functionality as our `widget-container` component, only it will use the defer control flow instead of the `ngComponentOutlet` directive. You can start by creating a component named `defer-widget`, next to the `widget-container` component. Once you've created the component. create a `widgets.enum.ts` file in the newly created `defer-widget` folder. Add the following enum inside the `widgets.enum.ts` file:

```
export enum Widgets {
  Clock,
  Weather
}
```

Next, you need to add an input for the active widget, the widget data, and a property with a reference to the `Widgets` enum inside `DeferWidgetComponent`:

```
@Input() activeWidget!: Widgets;
@Input() activeData!: any;
widgets = Widgets;
```

Inside the HTML template of `DeferWidgetComponent`, you need to add your widget components inside a defer block. Defer blocks receive a trigger as a parameter. When the defer trigger is triggered, the defer block will load and render the content that is inside the defer block. For your widget component, you need to use the when trigger. The when trigger loads and renders the content when the provided condition resolves to `true`.

We also want our content to hide again when the condition is `false` again. To hide your content when the condition is `false`, you also need to add a `*ngIf` directive or use the `if` control flow. I will use the `*ngIf` directive:

```
@defer (when activeWidget === widgets.Clock) {
  <bt-libs-clock-widget *ngIf="activeWidget === widgets.Clock" />
}
@defer (when activeWidget === widgets.Weather) {
  <bt-libs-weather-widget *ngIf="activeWidget === widgets.Weather"
[widgetData]="widgetData" />
}
```

Without the `*ngIf` directive or if control flow syntax, the widget will not disappear if the condition returns `false`. The when statement of the `defer` syntax will not evaluate the condition again after it has resolved to true.

The aforementioned code is everything you need to lazy load and render widgets. As you can see, this is a lot easier compared to using the `ngComponentOutlet` directive.

Because we already had some code in place, we still need to make some changes to the weather widget for our new defer solution to work. Inside the weather widget component, you have a `widgetData` property. You can place this under `comment` and replace it with a `@Input()` property, like this:

```
// widgetData: WeatherWidgetData = inject(WEATHERWIDGET);
@Input() widgetData!: WeatherWidgetData | null;
```

The `widgetData` property we commented out is used for the `ngComponentOutlet` directive approach. For our new defer approach, we will use a regular component, `@Input()`. You can also update the HTML template of the weather component and only render the city and message if we receive the `widgetData` property:

```
<p *ngIf="widgetData">{{widgetData.city}}: {{widgetData.message}}</p>
```

Now, everything should work! To test `DeferWidgetComponent`, we can use it in `expensesOverviewComponent`, where we now use `widgetContainerComponent`. In the HTML template, you can swap the old widget container with the new defer widget:

```
<bt-libs-defer-widget [activeWidget]="widget"
[widgetData]="widgetData" />
```

Now, we only need `widget` and `widgetData` properties inside `expensesOverviewComponent` and must set their values. You can comment out the old `widget` and `injector` properties we used for the widget container and add these two properties instead:

```
widget!: Widgets;
widgetData: any = null;
```

Lastly, we have to change the logic inside `setInterval`. Once again, you can comment out the old code we used to alternate the weather and clock widgets for the widgets container and replace it with this code:

```
this.widget = this.widget === Widgets.Clock ? Widgets.Weather :
Widgets.Clock;
this.widgetData = this.widget === Widgets.Clock ? null : { city:
'Amsterdam', message: 'Sunny' };
this.cd.detectChanges();
```

Now, when you save everything, the new defer widget component should work identically to the widget container component. After 5 seconds, the clock widget will be lazy loaded and rendered, and 5 seconds after that, the weather widget will be lazy loaded and rendered. The two widgets will alternate every 5 seconds.

With that, you've created a widget component with lazy loaded widgets using the ngComponentOutlet directive and the defer control flow syntax. With the ngComponentOutlet directive, you can provide injectors and you don't have to remove the widgets with the *ngIf directive, but overall, the defer control flow syntax feels a lot cleaner and easier.

Summary

We learned a lot in this chapter! First, you learned how to make components more flexible using content projection. We created a modal component and showcased content projection with a single slot and multiple slots. We also learned that you can't combine structural directives on projected content.

Next, we did a deep dive into template variables and template references. You learned how to create flexible and dynamic components using ng-template, how to access the values of components and input properties using template variables, and how to display different templates based on specific conditions. You also learned how to provide context to ng-template elements to build truly dynamic components that can fit almost all design needs you could have for a component.

Lastly, you learned about dynamic component rendering and loading. You learned when you should use dynamically rendered or loaded components and how you can dynamically render and lazy load components at runtime using the ngComponentOutlet and @defer syntax.

In the next chapter, we will start to learn about conventions and design patterns within Angular so that we can improve the setup and implementation of our code.

6

Applying Code Conventions and Design Patterns in Angular

In this chapter, we will explore code conventions, best practices, and design patterns commonly used within Angular applications. You will also create a generic HTTP service and mock API responses using an HTTP interceptor.

Following good code conventions allows you to write code consistently. Whether you write your code solo or in a team, conventions ensure that you use similar syntax for common occurrences and follow best practices. Using good design patterns helps you to write code implementations that scale well and are battle-tested.

Code conventions and best practices focus more on processes and style-related aspects such as using the CLI, naming, using types, or preventing nested observables. Design patterns, on the other hand, focus on how you set up, handle, and implement common occurrences, problems, and flows within your code base.

By the end of this chapter, you will know all about code conventions, best practices, and commonly used design patterns within Angular applications. Some patterns and principles you will learn about in this chapter are inheritance, facade services, observables, reactive programming, and anti-patterns. This chapter will provide a good foundation for the following chapters, where we dive deep into reactive programming and state management. The chapter will help you understand the benefits of good design patterns and code conventions.

This chapter will cover the following main topics:

- Exploring commonly used code conventions and best practices in Angular applications
- Exploring commonly used design patterns in Angular applications
- Building a generic HTTP service containing a model adapter

Exploring commonly used code conventions and best practices in Angular applications

In this chapter's first section, you will learn about **code conventions** and best practices within Angular applications. Using code conventions ensures that everyone working on your project uses similar naming for variables, files, and folders. Good code conventions also make code more readable and allow you to recognize certain features, implementations, or data types quickly. In addition, code conventions make your code consistent and easier to debug, refactor, and understand. Setting up good code conventions for your project promotes the usage of best practices. Code conventions also make it easier for new developers to be onboarded into the code base, as they have a set of rules they can follow that allows them to write code in a similar way to the rest of the people working on the code base.

Personally, I think it's a good practice to create a document with all code conventions and best practices you adopt within your project. This way, new people have something to go off of besides what they see in the code. What code conventions your company adopts is entirely up to the people writing and maintaining the code base. However, some commonly used conventions and best practices within the Angular community exist.

For starters, Angular has a *style guide* in which it declares everything it considers a good practice and why. You can find the style guide on the official Angular website at `https://angular.io/guide/styleguide`.

Next, we learn about common conventions and best practices for Angular applications. We will start with naming and structural conventions and follow this up with best practices.

Naming conventions

Naming is the main focus of conventions. **Naming conventions** are essential to ensuring maintainability and readability. Good naming conventions allow you to navigate the code base easily and find content quickly. Naming conventions apply to several aspects of the code, so we will divide them up, starting with folders and files.

Naming folders and files

File and folder names should clearly describe the intent of the folder or file. This way, you can quickly locate files and folders you need even when the project grows. For folders, you should use single words, but if you do use multiple words, you can separate them with a dash (`-`). For file names, you can use the format of `feature.type.ts`.

The feature describes what the file entails and the type refers to things such as components, services, directives, pipes, and so on (some examples include `expenses-list.component.ts`, `expenses.service.ts`, and `unit.directive.ts`). Use conventional names for the file types (`.component`, `.directive`, `.service`, `.pipe`, `.module`, `.directive`, `.store`, `.actions`, `.stories`).

For unit tests, use `.spec` for the type. Lastly, it's wise to prevent duplicate folder or file names. As your monorepo grows, avoiding duplicate file or folder names might not always be possible, but try to prevent it for as long as possible.

Besides naming conventions for files and folders, adhering to naming conventions within your code is vital. It would be best to keep naming consistent for your classes, properties, functions, selectors, and other aspects of your code. Having good naming conventions within your code helps you to quickly identify different parts of your code, improving readability and making refactoring and maintaining your code easier.

Naming conventions within your code

Angular is predominantly a class-based framework, so let's start by naming classes. All classes should use upper camel case. Upper camel case is when you start every word with an uppercase letter. The class names should equal the file feature combined with the file type. So, `expenses-list.component.ts` becomes `ExpensesListComponent` and `expenses.service.ts` becomes `ExpensesService`.

Another important part of Angular applications is the selectors of components, directives, and pipes. For components and directives, it's a convention to prefix the selectors. Make the selector prefix unique so you can distinguish it from selectors from any third-party libraries you might use.

Component selectors are all lowercase and words are separated with dashes. For directive selectors, you use regular camel case. With regular camel case, the first word is in lowercase and all subsequent words start with a capital letter. With selectors for pipes, you should use a single word in all lowercase without a prefix. For pipes, use regular camel case if you have to use multiple words for the pipe selector.

There are also some common conventions when we look at the code within classes (or function files):

- First, we use camel case to declare properties, functions, and methods. Using descriptive names for your properties, functions, and methods is also important. When you handle events or component outputs with functions, you should prefixed these functions with on (`onClick`, `onAddExpense`, `onHover`):

  ```
  <div (click)="onClick()"></div>
  ```

- When declaring a component output, you don't prefix it with on:

  ```
  @Output() saved = new EventEmitter<boolean>();    // Good
  @Output() onSaved = new EventEmitter<boolean>(); // Bad
  ```

- Lastly, when you declare observable properties, you suffix them with a dollar sign ($):

  ```
  import { interval } from 'rxjs'
  numbers$ = interval(1000);
  ```

Now that you know about naming conventions for files and naming conventions for your code, let's look at some conventions for structuring your files and projects.

Structural conventions

Besides naming conventions, you can have conventions for structuring your files and projects. Like good naming, having a predictable and good structure inside your files and projects helps with readability and maintainability. With a good file structure, you can easily recognize and find the parts of code that you need.

First, it's a good convention to use the **rule of one**. Each file should serve a single purpose. Having a single purpose for each file makes it easier to read and maintain them, and it keeps the files small. A single purpose for each file also makes it easy to locate bugs. It would be best if you also tried to limit files to a maximum of 400 lines. When a file exceeds 400 lines of code, it's a good indicator that you might need to split it up and move some methods to a separate file.

You should watch the size of your file and also the size of your functions. Ideally, functions should not be more than 50 lines, and it's best to keep them under 25 lines. Some exceptions might exist, but splitting them into separate functions when they grow larger is better. When functions grow too large, they become hard to read, test, and debug.

In Angular, you can write the template, CSS, and logic in a single file, but extracting the template and CSS into their own files is recommended. Using a separate file for the template and CSS promotes the rule of one, where each file has a single purpose, and it also helps with readability and maintainability. If your template consists of one or two HTML tags without additional styling, you can make an exception and place everything in a single file. However, I would still separate the template and component class. Besides separating your files into dedicated HTML, CSS, and TypeScript files, a good folder structure also helps with maintaining a clear overview, so let's look at some conventions for our folder structure.

Try to maintain a flat folder structure for as long as possible. Having a lot of nested folders makes it easier to find the folders and files you need and can make your overview of the folder and file structure clearer. Create a folder or, better yet, a library for each domain in your project. You should split your code inside these libraries into `data-access`, `features`, `UI`, and `utils`. Each element inside your `data-access`, `features`, `UI`, and `utils` libraries should also be its own library.

Using libraries promotes an API-driven architecture and ensures good separation of concerns. By having an API-driven architecture, you also start to write more reusable code. Inside each library, you have an `index.ts` file to export what you need to consume elsewhere.

Next, it's recommended to use the **DRY principle**. DRY stands for "Don't repeat yourself." When your mono repository grows larger, sometimes you can't help repeating yourself, but in general, you should try to write code only once and share it where you need it.

Lastly, you need a way to order your code within your files. A common way to order your code is by using the following structure:

1. `@Input()` decorators

2. `@Output()` decorators

3. Public properties and private properties

4. Constructor

5. Getters and setters

6. Lifecycle hooks

7. Public methods and private methods

Sort properties and methods alphabetically (after dividing them into public and private properties and methods). Initialize `@Input()` directives whenever possible, and when using lifecycle hooks, implement the interface as well.

You can extend the before-mentioned conventions to make your project more robust and uniform. The number of conventions you should come up with and try to use is unlimited, but for now, you have a good starting point and idea of what your conventions should look like and focus on. Next, we will discuss some best practices within Angular applications.

Using best practices in Angular applications

Using best practices ensures that you do things properly and your code stays robust, maintainable, and scalable.

The first best practice is to use the Angular CLI (or the Nx CLI when using Nx) as much as possible. Using the CLI to generate components, services, directives, projects, libraries, and other elements ensures consistency. When using the Nx CLI, you're also assured that all dependencies and settings are configured properly.

Use the new standalone components, directives, and pipes as much as possible. Using the new standalone API helps better isolate your logic, making debugging and testing your components, pipes, and directives easier. Using the standalone API also helps to reduce your bundle sizes, resulting in faster load times. You should also use the new `inject` function over constructor injection for dependency injection. The `inject` function provides more flexibility and is no hindrance when you use inheritance.

Always use access modifiers on your properties and methods. In Angular, we have three access modifiers: `public`, `private`, and `protected`. Using the correct access modifiers makes it easy to identify what can be used where and helps to prevent bugs and unintended behavior.

Performance-related best practices

There are a bunch of best practices related to performance. First, always use the `trackBy` function on the `*ngFor` directive. When you use the new control flow syntax introduced in Angular 17, you are required to use the `track` function. Using the new control flow syntax is recommended as it improves readability, and you don't need to import the common module to use them, so it will reduce your bundle sizes a bit.

Next, you should use *lazy loading* as much as possible as this ensures that you only download what your user requests. With the new standalone components, you can easily lazy load every route, and with new defer blocks introduced in Angular 17, you can even lazy load different parts of the HTML templates.

You should try to use the `OnPush` change detection strategy inside your components as much as possible. Using `OnPush` change detection reduces the number of times Angular renders your template. For even better change detection and performance, you should also utilize Angular signals as much as possible to manage the synchronous state in your application (we will discuss signals in detail in *Chapter 7*).

For asynchronous data flows, you should use **RxJS**. Observables should be handled with the `async` pipe as much as possible. The `async` pipe unsubscribes automatically for you when the component is destroyed or the property is assigned with a new observable; this prevents memory leaks and improves the performance of your application.

Don't use function calls or getters in your HTML templates. Calling functions or using getters in the template negatively impacts the performance of your application. Use **CDK virtual scroll** to display large lists. The CDK virtual scroll will only render the elements displayed inside the view instead of the entire list. Use pure pipes as much as possible:

```
<cdk-virtual-scroll-viewport itemSize="50" class="example-viewport">
   <div *cdkVirtualFor="let item of items" class="example-
item">{{item}}</div>
</cdk-virtual-scroll-viewport>
```

For example, when using the `ngOnChanges` lifecycle to assign a property based on the newly received input values, there is a big chance you can handle the same using a pipe. Using pure pipes is better for performance and promotes reusability. Don't use `filter`, `forEach`, `reduce`, or `map` on arrays inside pipes, as this negatively affects the performance.

Lastly, you should cache API requests for as long as feasible and do the same with resource-intensive methods.

Preventing bugs with best practices

Besides performance-related best practices, there are some best practices to prevent bugs and improve testability and maintainability. Avoid `any` types in your code. Having everything strongly typed prevents bugs, improves suggestions, and makes debugging and testing easier.

When you have observables inside component classes that aren't used within the template, use the RxJS `takeUntilDestroyed`, `takeUntil`, or `take` operators. These three operators ensure that your subscriptions on observables are unsubscribed correctly.

Also, don't use nested observables; instead, use RxJS operators such as `combineLatest` and `withLatestFrom` to handle scenarios where you need nested observables. Nested observables can quickly lead to memory leaks and bugs that are difficult to debug. Nested observables are also hard to write tests for.

Additionally, avoid using multiple `ng-container` elements with an `async` pipe when you must await multiple observables before rendering a piece of HTML.

Don't do this:

```
<ng-container *ngIf="obs$ | async as observable">
  <ng-container *ngIf="obs2$ | async as observable2">
  </ng-container>
</ng-container>
```

Instead, map both observables into a single observable inside your component class and use the single observable in the template with the `async` pipe:

```
observables$ = combineLatest({a: of(123), b: of(456)}).pipe(map(({a,
b}) => ({obs: a, obs2: b})));
```

Combining the observables will prevent bugs, make your template more readable, and ensure everything is updated correctly when an observable receives new values.

Now that you know some best practices that prevent bugs, let's explore best practices regarding your setup and architecture.

Best practices for your project setup and architecture

A good setup helps with better maintainability and standards throughout your codebase. We already discussed how to set up a project in *Chapter 1*, but to recap, use smart and dumb components.

Smart components connect with your state management and *dumb components* only receive data through inputs and output changes to the parents. This ensures you don't have unintended dependencies and your component focuses on a single responsibility. Use `export default` on your components as much as possible to auto unwrap when lazy loading. Using default exports with automated unwraps keeps your routing files clean and readable.

Use the `canMatch` route guard over the `canActivate` and `canLoad` guards. The `canMatch` guard will not download the code if the guard returns false. Lastly, you need to use lint rules to enforce your conventions and best practices.

You have learned about naming and structural conventions. You also learned about best practices you can use within your Angular applications. Going forward, I will mention other best practices as we go, but first, we will learn about common design patterns in Angular applications.

Exploring commonly used design patterns in Angular applications

Design patterns help solve common software development problems in a predefined approach. Design patterns are like blueprints for building your application. Design patterns tell you how your code should behave and how to create a structure or separate parts of your code base. Using design patterns ensures that you have battle-tested solutions for common problems with a good level of abstraction so your code can scale without becoming a mess, entangled with dependencies all over the place, which is commonly referred to as "spaghetti code."

Abstraction in software development means you separate the details and behavior of your system from the implementation logic. For example, if you have a state management solution in your Angular application, you should separate the implementation of your state management from your component layer. By separating the component layer and state management solution, you can change your state management solution without touching your component layer. This provides extra flexibility and can save you some severe refactoring down the road as your application grows and its needs change.

To stay with the state management example, in the beginning, you might manage your state using a simple approach with RxJS `Subject` and `BehaviorSubject`, but when the application grows and your state becomes more complex, you might want to change it for something such as **NgRx** or **NgXs** as they offer a safer, more robust and flexible approach for handling complex application state.

Suppose your component layer is entangled with your state management. In that case, you have to refactor your entire application to switch the state management implementation. In contrast, if you have a good level of abstraction between the component layer and the state management solution, you can change the state management implementation without touching your components.

Design patterns are a good starting point for solving common problems in software development, but they are no holy grail or a one-size-fits-all solution. You should always consider what is useful for your application; don't overdo it by using design patterns where none are needed. When it makes sense to follow design patterns to the letter, you can do so, but when it doesn't, adapt the patterns to fit your specific needs.

Now, without further ado, let's explore some commonly used design patterns in Angular applications.

Creational design patterns in Angular

Creational design patterns form the foundation of how we create classes and objects within our applications. Within Angular applications, creational patterns are used by the framework for the creation of components, services, and other essential building blocks of the application. Developers can ensure modular, reusable, and maintainable code by implementing creational patterns, such as factory and singleton patterns.

Singleton pattern

The **singleton pattern** is used to create a single instance of an object or class. Using the singleton pattern ensures that all code interacting with the singleton uses the same instance. Another advantage of the singleton pattern is good memory usage, as you only have to allocate memory for the object or the class once. In *Figure 6.1*, you can see a visual representation of the singleton pattern:

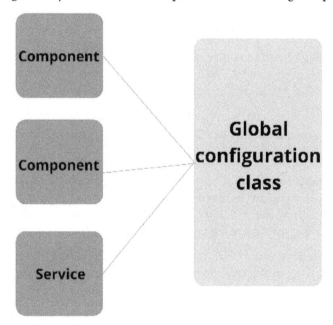

Figure 6.1: Singleton pattern

As you can see in *Figure 6.1*, there is one single instance of a class used by different consumers. When you don't use the singleton pattern, you have multiple instances of the global configuration class, where each consumer uses its own instance. Now that you know what the singleton pattern is, let's explore how the pattern is commonly used in the context of Angular.

Within the context of Angular, the singleton pattern is mostly used in combination with dependency injection. When you create a service or provide other dependencies in your Angular applications, it's generally done as a singleton, meaning only one instance of the dependency is created and shared by all consumers throughout the application. Because there is only one instance, you can safely use singleton services and classes to manage the state within your Angular application or handle other logic where each consumer needs access to the same instance, such as configurations and caching.

To provide a dependency as a singleton, you must provide it within your application's *root providers array*. When working on an Angular application without modules, the root providers array is located in the `ApplicationConfig` object you provide to the `bootstrapApplication` method. If you use `ngModules` in your Angular application, you make a dependency a singleton by providing it in the app module. You can also use the `providedIn` root configuration object when it comes to services:

```
@Injectable({ providedIn: 'root'})
export class ExpensesService {}
```

In general, you can really only create a single instance of objects utilizing the singleton pattern. In the case of Angular dependencies, a singleton is created within the context of the provider array where the dependency is provided. We already explained the providers array and creation of dependencies in more detail in *Chapter 2*.

To further clarify the singleton pattern, let's look at some real-world examples of when the pattern is useful:

- **Managing the logged-in user**: If you have a class to manage the logged-in user, you want there to be a single instance of that class so that there is a single source of truth. If there are multiple copies of the class, the user data, login state, and other properties could vary amongst the different instances of the class, resulting in unintended behavior.

- **State management**: When you have a class to manage the global application state, the singleton pattern is also a good fit. You want to ensure that everyone who needs the global application state receives the same value and can update it in the same source. If there are many instances of the state class, these instances can hold different values, resulting in a corrupted state. Keeping different instances synchronized can be a hard task, so using a single source of truth makes sense and uses less memory.

Now that you know about the singleton pattern and when it's used within the Angular application, let's explore the factory pattern.

Factory pattern

The **factory pattern** serves as a versatile blueprint for object creation. The factory pattern is beneficial in scenarios where the exact type of object needed is determined at runtime, allowing for dynamic instantiation based on certain conditions or parameters. By encapsulating object creation, the factory pattern prevents tight coupling between client code and specific classes, promoting maintainability, scalability, and easier modifications. *Figure 6.2* shows a visual representation of the factory pattern:

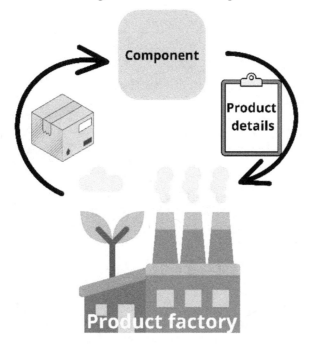

Figure 6.2: Factory pattern

As you can see in *Figure 6.2*, we send the factory some information (in the figure, it's a list of product details) and the factory creates what we want it to create and returns the result. In the figure, the factory creates and returns a product to the component using the factory. Now that you have a better idea of what the factory pattern is, let's see how it is commonly used in the context of an Angular application.

Within the context of Angular, the factory pattern is also mainly used in combination with dependency injection. You can create a provider with the `useFactory` property and provide a factory method for creating the dependency value. Using the `useFactory` property in a provider can be useful when you want to provide different classes based on a condition or when you want to provide a value to the created class that is only accessible upon runtime:

```
providers: [
   { provide: LoggerService, useFactory: env.prod ? ProdLogger :
DevLogger }
]
```

Another place where you commonly use factory methods within Angular applications is in services. Oftentimes, services are used to create specific objects; this can be done with a factory method so you have a concise way to create the objects.

```
createProduct(name: string, props: ProductProps): Product {
   return new Product(name, props);
}
```

Now you know the factory pattern is used to create objects and classes in a predictable manner. Next, we will talk about the dependency injection pattern.

Dependency injection pattern

Dependency injection (**DI**) stands at the core of the Angular framework. Strictly speaking, DI falls under the umbrella of the **inversion of control** (**IoC**) principle. IoC essentially delegates control of certain aspects of a program to an external framework or container, allowing it to manage the flow and connections between components. But because DI is in charge of the creation and distribution of dependencies within your Angular application, we can place it under the category of creational design patterns.

DI promotes modularity by decoupling components and services, making them more reusable across the application. By facilitating the management of dependencies, DI makes it easier to scale applications by adding or modifying functionality without needing to make significant code changes.

The Angular DI system helps identify and prevent circular dependencies, which can lead to runtime errors and hard-to-debug issues. Dependency injection also enforces type safety, reducing the risk of errors related to incorrect data types being injected.

We already discussed Angular DI in great detail in *Chapter 2*, so for now we will move on to structural design patterns.

Structural design patterns in Angular

Structural design patterns are fundamental in shaping the architecture of your Angular applications. Structural design patterns help organize components, services, and modules, defining how they interact and collaborate within your applications. Angular leverages these patterns to establish a clear structure, facilitating the development of scalable, modular, and maintainable applications.

For instance, patterns such as component-based architecture and module structure are inherent in Angular applications. The decorator pattern is also heavily used in Angular (e.g., `@Component`, `@Injectable`). The facade pattern is often used in Angular applications to provide abstraction between the services and component layer of your Angular applications.

Overall, using structural design patterns in Angular guides developers in creating well-organized, scalable, and adaptable applications by defining how components, services, and modules interconnect and collaborate within the framework's architecture.

Component-based architecture

Component-based architecture (CBA) is a design pattern where we build applications by composing individual, self-contained, and reusable components. It's obvious where CBA comes into play within Angular applications. When building components, it's important to keep them as self-contained and reusable as possible. Because we want to build reusable components, it's important to use the smart/dumb principle within your Angular components.

We already talked about smart and dumb components in *Chapter 1* and *Chapter 2*, but to reiterate, smart components are mainly pages or large feature components and have a connection with your business logic and state management (or the facade services).

Dumb components are UI elements that are used to build up the smart components. Dumb components receive their data through inputs and notify other components of changes with outputs. Using this smart/dumb approach enforces good architecture and reusability of your components.

Now that we have briefly reiterated the CBA pattern, let's move on and learn about the decorator pattern.

Decorator pattern

The **decorator pattern** is a structural design pattern that enables you to modify the behavior of classes, functions, and properties without altering the object itself. The most commonly used decorators within the Angular framework are as follows:

- `@Component`: This decorates a class as an Angular component, providing metadata for Angular's compiler. It includes information about the component's template, styles, and other configurations such as the standalone flag, imports, the component selector, and directive decomposition.

- `@Injectable`: This decorates a class as an injectable service, allowing it to be injected into other components or services through Angular's DI system.

- @NgModule: This decorates a class as an Angular module, providing metadata that defines the module's dependencies, components, directives, services, etc.

- @Input and @Output: These are decorators used in component properties to define inputs and outputs for communication between components.

- @HostListener: This decorates a class method to declare a DOM event listener. It's used within directives to listen for events on the host element.

- @HostBinding: This decorates a class property to bind it to a host element property within a directive.

- @ViewChild and @ViewChildren: These are decorators used to query and obtain references to child components or DOM elements within a parent component or directive.

You can also create your own custom decorators. We can, for example, make a custom decorator that will log when a function is called and include the provided function parameters in the log. To start, you need to set experimentalDecorators to true inside your tsconfig.json:

```
{
    "compilerOptions": { "experimentalDecorators": true}
}
```

In the case of Angular, the experimentalDecorators property is set to true by default because Angular already uses decorators within the framework. To create a decorator, you have to create a function that takes three arguments: target (class prototype), propertyKey (name of the method), and descriptor (property descriptor of the method). Within the decorator, you modify the behavior of the decorated method by wrapping its original logic inside a new function. This new function logs a message before invoking the original method:

```
export function LogMethod(target: unknown, propertyKey: string,
descriptor: PropertyDescriptor) {
  const originalMethod = descriptor.value;

  descriptor.value = function (...args: unknown[]) {
    console.log(`Method ${propertyKey} is called with args: ${JSON.
stringify(args)}`);
    return originalMethod.apply(this, args);
  };
  return descriptor;
}
```

Now, to use the decorator, you simply add it above a method, like this:

```
@LogMethod
test(a: number, b: number) {
```

```
    return a + b;
}
```

If you now call the test method, the decorator makes sure it is logged, including the a and b parameters:

```
this.test(1, 2);
Logs: Method test is called with args: [1,2]
```

You can place the custom decorator in a new custom-decorators library of type util in the shared domain. I created a func-logger.decorator.ts file inside the custom-decorators library and placed the logic of the decorator in that file.

You now know where Angular uses the decorator pattern and how you can use it yourself to extend or modify the behavior of objects, functions, and classes without modifying the objects themselves. Next, you will learn about the facade pattern.

Facade pattern

We mentioned the **facade pattern** a couple of times throughout this book. Now it's time to explain what the facade pattern is. The facade pattern is a structural design pattern that provides a simplified interface to a larger, more complex system of classes, subsystems, or APIs. In the context of Angular applications, the facade pattern is commonly used to create a simple interface for and an abstraction layer between the component layers and your services where you implement your state management solution and business logic. In *Figure 6.3*, you'll find a visual representation of the facade pattern:

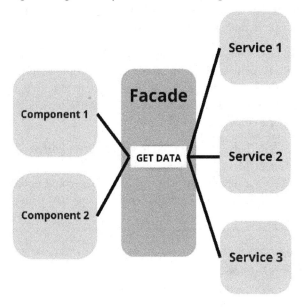

Figure 6.3: Facade pattern

In *Figure 6.3*, to get the data needed in the component, three different services need to be called. If you have to do this for all your components, you create a lot of dependencies and you need to create a lot of logic inside your components. Instead of directly calling the services from the components, we place a facade in between the components and services. The facade provides a simple `get data` method to retrieve the data we need inside the components. Instead of having dependencies inside the components, the facade has all the dependencies to the services. By using the façade, you can keep your components simple and clean without dependencies. Furthermore, you create a level of abstraction, and because of that, you can change the how you retrieve the data without changing the component layer; you simply change how the facade gets the data and the component still calls the `get data` method from the facade service. In most cases, you have your state management on the right side of the facade and your components on the left side of the facade.

In large applications and monorepos, the state management layer can become large and complex to handle. For something as simple as getting all expenses in a large monorepo of business tool applications, you can find yourself accessing multiple state files (commonly named stores) and selectors of these stores. Because the pieces of the state can live in different files or even different libraries, it's not as easy as creating a function that combines the parts inside the state itself. You need another place, and that would be the facade service. The same goes for updating; in large systems, this can involve updating multiple stores and handling multiple callbacks or effects.

Let's say you create an expenses facade. In this facade, you create methods that are simple to call and give you access to everything you need in your component layer. Think of things such as getting all expenses, getting filtered expenses, updating a single expense, or updating expenses in bulk. The facade helps keep things simple in your component layers and ensures that you only have to call a single function to get or do what you need to make your components work and update your state accordingly.

The facade makes accessing and updating your state easy and provides an extra level of abstraction, decoupling your state management implementation from your component layer. This abstraction layer allows you to change your state management solution without touching your component layer. You only have to change your state management and update the facade and your component layer will keep working as if nothing changes.

If you do not have the facade in between, you need to go into every component where you use the state and update them all independently, resulting in more work and a higher chance of you missing something. So, even if you have a simple state where you can access and update most things with a single method, adding a facade between your component layer and state management is still wise.

The following is a simple example of how a facade service might look:

```
@Injectable({ providedIn: 'root' })
export class ExpenseFacadeService {
  protected readonly expenses: inject(ExpensesStore);
  protected readonly approvedExpenses: inject(ApprovedExpensesStore);

  getAllExpenses(): Expense[] {
```

```
      return [...this.expenses.getAll(), ...this.approvedExpenses.
getAll()];
  }
  addExpense(expense: Expense): void {
    this.expenses.addExpense(new Expense(expense));
  }
  updateExpense(expense: Expense): void {
    if (expense.isApproved) {
      this.approvedExpenses.addExpense(expense);
    } else {
      this.expenses.addExpense(expense);
    }
  }
}
}
```

In the preceding example, we made a façade service that exposes two methods: one to get all expenses and one to update expenses. If the implementation in the state management now changes, you only have to change it in the facade instead of everywhere you get or update the expenses in your components. The preceding code is just an example; you don't have to add this in the monorepo. In *Chapter 8*, we will create a facade in our monorepo. For now, you just need to know what the facade pattern is and why it's useful.

Next, we will take a look at the inheritance pattern.

Model adapter pattern

The **model adapter pattern** is an implementation of the adapter pattern. The model adapter pattern is commonly used to map objects received from the API to a representation of those models used within the frontend (not specifically into models used within the view layer). You might ask yourself, why is this useful?

Imagine you receive an object from the API with a `title` property:

```
{
  title: "My Title",
  ......
}
```

Let's say you use this `title` property in 100 different places throughout your application. If, for some reason, the backend has to change the property name from `title` to `subject`, you need to go to all 100 places within your application to change them from `title` to `subject`. If you have a model adapter, you can change them in one place and map the new `subject` property to the `title` property of your frontend model.

Inheritance pattern

The **inheritance pattern**, a foundational concept in object-oriented programming, establishes a hierarchical relationship between classes, enabling one class (the subclass or derived class) to inherit properties, methods, and behaviors from another class (the superclass or base class). In Angular applications, services and components can utilize inheritance to form hierarchical relationships and share generic functionalities and properties. The base class exposes the shared functionalities and properties with the child class.

Inheritance is a powerful and useful design pattern, but it should be used in moderation, especially within the component layer of your Angular applications. Excessive use of inheritance can create tight coupling between the base and child components. Changes in the base component might inadvertently impact multiple derived components, making the system fragile and harder to maintain. Deep hierarchies with multiple levels of inheritance can introduce complexity, making the codebase harder to understand and maintain. Over-engineering by creating overly complex inheritance structures might hinder rather than aid development.

Also, in the context of the component layer, you can often create pipes and directives to share common functionalities. For example, you can have a base class where you add common logic for handling disabled states, handling some commonly used component stylings such as `primary`, `alert`, and `danger`, and adding the option to change this styling when clicked or double-clicked.

This might seem like a valid solution, but you might only need one or two of the options in some scenarios. Some components might have different styling types but should not be able to change the styling type based on a click, and others should not be able to be disabled. It's generally a bad practice to expose behavior to a component that does not apply to the component. So, in this scenario, it would be better to create three directives that handle disabling, styling, and styling change behaviors and apply them using directive composition.

So, inheritance can be used to share common functionalities, but make sure it is the right solution for the problem so that you don't create base classes that expose a lot of functionalities and behaviors that most child classes will not use. One way I like to use the inheritance pattern is for creating a generic HTTP service. We will create a generic HTTP service with a model adapter in the next section.

Now that you have learned about the most used structural design patterns within Angular applications, we will start learning about common behavioral design patterns within Angular applications.

Behavioral design patterns in Angular

Behavioral design patterns focus on how objects and classes communicate and delegate responsibilities to each other. Using behavioral patterns ensures that your code remains flexible, modular, and maintainable.

The most common behavioral pattern within Angular applications is the observer pattern, but others, such as the interceptor, redux, and strategy patterns, are also commonly used. Let's start with the most commonly used: the observer pattern.

Observer pattern

If you have ever used Angular, you know the framework heavily relies on **observables**. The Angular framework integrated RxJS to manage observables and handle asynchronous data streams effectively and in a reactive manner. RxJS is a library focused on handling observable data streams. The **observer pattern** allows you to create one-to-many relationships so that when one object changes, all the dependent elements within your code are notified and updated automatically.

Within the observer pattern, you have the **observable** and the **observers**, which are more commonly referred to as the subscribers. I like to explain the Observer pattern with a magazine subscription analogy.

Let's say there is a magazine that releases a new issue weekly. The magazine is the observable and the people subscribing to the magazine are the observers or subscribers. Each time the magazine releases a new issue, all subscribers are notified and receive the new issue automatically in their mailboxes. The subscribers to the magazine will receive the magazine for as long as they are subscribed, and when they unsubscribe, they will no longer receive the magazine.

The magazine can be subscribed to by all people who want to receive it. There is one magazine issue and many readers of the issue, hence the one-to-many relationship, as we have in the observer pattern. There are also observers and an observable, just as with the observer pattern.

Observables come in two types, hot and cold observables, and within the context of Angular applications and RxJS, there are a couple of ways you can set up observables; we will dive deeper into observable types, RxJS, and handling observable streams in *Chapter 7*.

Interceptor pattern

The **interceptor pattern** allows you to intercept communications and perform some logic on the data that is being transferred and either stop or continue the intercepted communication.

Within the context of Angular applications, you commonly see the interceptor pattern in two different places:

- **Route guards** intercept route changes and either allow or block the route change based on some logic.
- **HTTP interceptors** intercept HTTP requests and responses. The HTTP interceptor is commonly used to add authorization headers to HTTP requests and handle retry logic or logging.

We will create a route guard in *Chapter 9*. For now, we will only create an HTTP interceptor.

Creating an HTTP interceptor can be done through a class-based or functional approach. We will use the functional approach, as this is the newer method and requires less boilerplate code.

To create an HTTP interceptor, you first need to make some adjustments to your `ApplicationConfig` object. Start by adding the `provideHttpClient` function inside your `providers` array:

```
providers: [provideHttpClient(), ...... ]
```

Next, inside the `provideHttpClient` function, add the `withInterceptors` function and provide it with an array:

```
provideHttpClient(withInterceptors([]))
```

Inside the array, within the `withInterceptors` function, you will register your HTTP interceptors. You create an interceptor by creating a function that implements the `HttpInterceptorFn` interface. The function takes an `HttpRequest` and an `HttpHandlerFn` as function parameters. The `HttpRequest` gives you access to the request and the `HttpHandlerFn` is called to continue the request after you perform your logic:

```
export const AuthInterceptor: HttpInterceptorFn = (
  req: HttpRequest<unknown>, next: HttpHandlerFn) =>  next(req.clone({
setHeaders: { Authorization: 'auth_token' } }));
```

The preceding code is just a simple example; we add a string as our authorization token. In reality, you should store your token in a safe place, such as in your environment variables, and retrieve it from there. You can also add more logic to the interceptor; this is just a simple example to illustrate how you create an interceptor. To activate the interceptor, you need to add it to the array of the `withInterceptors` function:

```
provideHttpClient(withInterceptors([AuthInterceptor]))
```

Now the interceptor is active and an authorization header with a value of `auth_token` will be added to all your HTTP requests. Now that you know what the interceptor pattern is and how you can create functional HTTP interceptors, let's move on to the Redux pattern.

Redux pattern

The **Redux pattern** is another commonly used design pattern within Angular applications. When you think about the Redux pattern, you might think of the Redux library, commonly used within the React framework, but the Redux pattern and the Redux library are two different things. The Redux pattern is a design pattern implemented by multiple state management libraries. Within Angular, the Redux pattern is commonly implemented by using the *NgRx* or *NgXs* state management libraries.

The Redux pattern focuses on predictable state management where the entire application state is stored in a single immutable state tree. The core principles of Redux involve defining actions that describe state changes and reducers that specify how those actions modify the state. The Redux pattern enforces a unidirectional data flow, allowing data changes to flow in a single direction, from actions to reducers to updating the application state. Changes to the state are made with pure functions called reducers (pure functions are functions that have the same output given the same input without performing any side effects). Retrieving the state is done by using pure functions named selectors.

In Angular applications, we create reactive code using observable streams. Because Angular uses observable streams, the libraries implementing the Redux pattern for Angular combine it with RxJS so that we can use the Redux Pattern in a reactive manner. This means actions are dispatched asynchronously, and you can use RxJS pipe operators in combination with the selectors for your state.

The Redux pattern is a bit too complex to explain in a couple of paragraphs, so we will come back to this topic in *Chapter 8*. For now, you just need to know that the Redux pattern focuses on handling state in an immutable and unidirectional way and has four key elements:

- **Actions**: It describes unique events to modify the state.
- **Reducers**: It has pure function implementations to modify the state based on the described actions.
- **Selectors**: It has a pure function to retrieve pieces of the state.
- **Store**: It has a class defining the state.

You now know about creational, structural, and behavioral design patterns within Angular applications. You learned about patterns such as the singleton, factory, decorator, facade, observer, and Redux patterns. You learned when these patterns are used by the framework and how you can use them to improve your code.

Building a generic HTTP service containing a model adapter

To build your generic HTTP service with a model adapter, start by creating a new library named `generic-http of type data-access` in the domain shared. In this library, create a file named `generic-http.service.ts` and a file named `model-adapter.interface.ts`.

Inside the `model-adapter` interface file, add this interface:

```
export interface ModelAdapter<T, S> {
  fromDto(dto: T): S;
  toDto(model: S): T;
}
```

We use generic types so we can make our model adapter type-safe. The T and S are placeholders for the **data transfer object (DTO)** and frontend model we will provide to the adapter. After creating the interface, start with the generic HTTP service.

The generic HTTP service will need a property for the API URL and the default HTTP headers, and the service needs to inject the HTTP client.

```
@Injectable({
  providedIn: 'root'
})
export abstract class GenericHttpService<T, S> {
  protected url;
  defaultHeaders = new HttpHeaders();
  protected readonly httpClient = inject(HttpClient);
}
```

As you can see, we also use generic types in the model adapter so we can maintain a type-safe generic HTTP service. For the HTTP service, we will also use the T and S as placeholders for the generic types. Next, we add the constructor so we can pass an API endpoint, base URL, and model adapter to the generic HTTP service:

```
constructor(
  private endpoint: string,
  private baseUrl: string,
  private adapter: ModelAdapter<T, S>
) {

  this.url = this.baseUrl + '/api' + this.endpoint;
}
```

Inside the constructor function brackets, we will construct the API URL from the base URL and the endpoint we receive from the constructor parameters. Now, you need to implement the API requests. The generic HTTP service will be used for common API requests: get, get by id, post, put, patch, and delete. Each request will use the URL we constructed, so this approach will only work if your API shares the same API route for these requests. Otherwise, you need to add an additional parameter to your constructor for the API routes (this can be done with a Record class, for example).

The API requests will include the request headers, and there will be an option to append additional headers if needed. Each request will also implement the model adapter so the objects received from the API will automatically be adapted to the frontend models. Here is an example of the get and post request:

```
public get(extraHttpRequestParams?: Partial<HttpHeaders>):
Observable<S[]> {
```

```
    return this.httpClient.get<T[]>(`${this.url}`, this.
prepareRequestOptions(extraHttpRequestParams)).pipe(
        map((data: T[]) => data.map(item => this.adapter.fromDto(item)
as S)));
}


public post(body: S, extraHttpRequestParams?: Partial<HttpHeaders>):
Observable<S> {
        return this.httpClient.post(`${this.url}`, this.adapter.
toDto(body), this.prepareRequestOptions(extraHttpRequestParams)).
pipe(map(data => this.adapter.fromDto(data as T) as S)) as
Observable<S>;
}
```

You can implement the other requests by yourself or take them from the GitHub repository from this book.

As you can see, the requests implement a function `prepareRequestOptions`. This `prepare RequestOptions` function is used to append additional API headers if needed. The implementation for this function looks as follows:

```
public prepareRequestOptions(extraHttpRequestParams = {}) {
   return {
      headers: Object.assign(this.defaultHeaders,
extraHttpRequestParams)
   };
}
```

Now you can add additional HTTP headers if needed. Now that you've created a generic HTTP service, let's explore how you can use the service.

Using the generic HTTP service

Using the generic HTTP service is done with inheritance when you create other HTTP services. For example, we need an HTTP service in our finance `data-access` library to get, update, and delete our expenses. Suppose you're not using the generic HTTP service with a model adapter. In that case, you will create an HTTP file in the finance `data-access` library that would look similar to the generic HTTP file, only with predefined types API URLs and models.

Because all these HTTP services commonly look more or less the same, with the exception of the API URL and the models, we created the generic HTTP service so we don't have to rewrite the same every time. Instead, we can inherit the generic HTTP service and share the common functionality.

Let's implement the generic HTTP service and start by creating a new folder named HTTP inside the finance data-access library in your *Nx monorepo*. Inside this new HTTP folder, create a file named expenses.http.ts. In the expenses.http.ts file, you have to create an ExpensesHttpService like this:

```
@Injectable({ providedIn: 'root' })
export class ExpensesHttpService{}
```

Next, create an interface for the expense model and expense DTO. The DTO represents the object structure we receive from the API and the model of the object structure we use in the frontend. I created a new folder named models inside the lib folder of the finance data-access library and added a expenses.interfaces.ts file in the models folder. In this expenses.interfaces.ts file, I will create the expense model and DTO interfaces:

```
export interface ExpenseDto {
  id: number | null;
  title: string;
  amount: number;
  vatPercentage: number;    date: string;
  tags?: string[];

}
export interface ExpenseModel {
  id: number | null;
  description: string;
  amount: {
    amountExclVat: number;
    vatPercentage: number;
  };
  date: string;
  tags?: string[];
}
```

Next, you can create a new folder named adapters. The adapters folder is located in the same place as the models and HTTP folders. Inside the adapters folder, create an expense.adapter.ts file, which will contain the model adapter for the expense DTO and model:

```
export class ExpensesModelAdapter implements ModelAdapter<ExpenseDto,
ExpenseModel> {
  fromDto(dto: ExpenseDto): ExpenseModel {
    return {
      description: dto.title,
      amount: {
        amountExclVat: dto.amount,
        vatPercentage: dto.vatPercentage
```

```
      },
      date: dto.date,
      tags: dto.tags,
      id: dto.id
    };
  }


  toDto(model: ExpenseModel): ExpenseDto {
    return {
      id: model.id ? model.id : null,
      title: model.description,
      amount: model.amount.amountExclVat,
      vatPercentage: model.amount.vatPercentage,
      date: model.date,
      tags: model.tags ? model.tags : []
    };
  }
}
```

As you can see, the model adapter maps the expense DTO to the expense model and the expense model to the expense DTO. We will provide this model adapter to the constructor of the generic HTTP service so that our models will automatically be mapped when we receive them from or send them to the API. The last step is to inherit the generic HTTP service in your expenses HTTP service. You inherit by using the extends keyword and adding the class name of the generic HTTP service:

```
@Injectable({
  providedIn: 'root'
})
export class ExpensesHttpService extends
GenericHttpService<ExpenseDto, ExpenseModel> {
  constructor() {
    super(
      '/expenses',
      <>,
      new ExpensesModelAdapter()
    );
  }
}
```

As you can see, we also call a super() method and provide it with some arguments. The super() method is used to call the constructor method of the inherited class – in our case, GenericHttpService. You provide the super() method with the properties that the constructor of the GenericHttpService expects to receive, the endpoint, baseUrl, and model adapter.

You might also notice we use the `arrow` syntax after the `GenericHttpService` and provide the `ExpenseDto` and `ExpenseModel` inside:

```
GenericHttpService<ExpenseDto, ExpenseModel>
```

Using the before-mentioned syntax provides the generic HTTP service with the generic types we added in the `GenericHttpService` class:

```
GenericHttpService<T, S>
```

By using generic types, we make sure the generic HTTP service remains type-safe, and we receive typed objects when we use the methods of our HTTP services. After inheriting the generic HTTP service inside the expenses HTTP service, you are ready to use the expenses HTTP service. All the methods, such as `get`, `get by id`, `post`, `delete`, and so forth, are inherited by the generic HTTP service and don't have to be implemented again.

If you want to use the expenses HTTP service, you `inject` it as you would with any other HTTP service or dependency:

```
protected readonly expensesApi = inject(ExpensesHttpService);
```

After injecting the HTTP service, you can call any method the service exposes:

```
this.expensesApi.get()
  .pipe(takeUntilDestroyed(this.destroyRef))
  .subscribe((data) => { console.log(<data ==>>, data); });
```

When we make the `get` request, you might notice the request is failing; this is because we don't have an actual API running and the request to `/api/expenses` returns a `404 not found` error code. Let's resolve this issue and provide some mock data when we make our API requests.

Providing mock data for our API requests

There are many ways to provide mock data; we will use an HTTP interceptor to get our mock data from the assets folder. This is just a simple implementation, and it has some flaws, but for demonstration purposes, it works very well, and it doesn't take much effort to set up.

We start by creating an interceptors folder with a `mock.interceptor.ts` file inside the `generic-http` library in our Nx monorepo. Inside the `mock.interceptor.ts` file, we will create an interceptor that returns the HTTP request untouched if we are not in development mode.

If we are in development mode, the interceptor will adjust the request URL and request method so that all requests are GET requests, and it will try to get a JSON file to place in the `assets` folder of our applications (in our case, the *expenses-registration application*).

We will also intercept the HTTP response, and if we do not make a GET request, we will return the request body instead of the data from our JSON file in the assets folder of the project.

```
export const MockInterceptor: HttpInterceptorFn = (
  req: HttpRequest<unknown>,
  next: HttpHandlerFn,
) => {
  if (!isDevMode()) return next(req);

  const clonedRequest = req.clone({
    url: `assets${req.url}.json`,
    method: <GET>,
  });

  return next(clonedRequest).pipe(
    map((event: HttpEvent<unknown>) => {
      if (event instanceof HttpResponse && req.method !== 'GET') {
        // Modify the response body here
        return event.clone({ body: req.body });
      }
      return event;
    }));
};
```

As you can see, we first check that we are not in development mode, and if that is the case, we directly return the request. If we are in development mode, we clone the request and adjust the request URL and method. The method is set to a GET, regardless of what request we try to make. The URL is prefixed with assets, so we target the assets folder of our application and postfix it with .json because we will fetch a JSON file from the assets folder.

Adjusting the URL in this manner makes sure that a request to the following API URL /api/expenses will be transformed to /assets/api/expenses.json. Next, we use the RxJS pipe and map operator to listen for the HTTP response and simply return the request body if the original request method wasn't a GET request.

Next, you have to create an api folder inside the assets folder of the *expenses-registration application*, and in this api folder, you need to create an expenses.json file with your mock data:

```
[
  {
    "id": 1,
    "title": "Office Supplies",
    "amount": 50.0,
    "vatPercentage": 20,
    "date": "2019-01-04",
```

```
    "tags": [
       "printer"
    ]
  },
  ..........
]
```

Lastly, you need to register the interceptor in your `ApplicationConfig` object:

```
provideHttpClient(withInterceptors([MockInterceptor])),
```

After adding the interceptor in the `ApplicationConfig`, your API request will return with your mock data for `GET` requests and the request body for all other types of requests.

If you make API requests to other API endpoints, you need to add additional JSON files in the `assets` folder to mock the response for these requests. That was it for the generic HTTP service and model adapter. In the next chapter, we will start implementing RxJS and signals and use the observer pattern in practice.

Summary

In this chapter, you learned about Angular code conventions and best practices. You learned about naming conventions for your folders and files and naming conventions for properties, functions, and classes within your Angular applications. You also learned about best practices to improve performance and prevent bugs and hard-to-debug code.

Besides conventions and best practices, we looked at some of the most commonly used design patterns within Angular applications. You learned when the Angular framework uses specific patterns and how you can write cleaner and more scalable code using design patterns such as the facade, decorator, and inheritance patterns.

We finished the chapter by creating a generic HTTP service using the inheritance pattern. The HTTP service can easily be used to construct HTTP services for all your data access libraries. The generic HTTP service also has a model adapter to transform DTOs into your frontend models automatically. Lastly, because we don't have an API, we made an HTTP interceptor to intercept our HTTP requests and provide mock data.

In the next chapter, we will learn about reactive programming and implement RxJS and signals within our expenses application.

7

Mastering Reactive Programming in Angular

Reactive programming helps improve your applications' performance and allows Angular to make better use of the change detection mechanism, reducing the number of times your application needs to be re-rendered.

In this chapter, you'll learn about reactive programming. You will learn what reactive programming is and how it can be used to improve your Angular applications. You will also learn about the RxJS library and how it can manage asynchronous data streams reactively. This chapter will teach you how to use different RxJS operators, create reusable RxJS operators, reuse sets of RxJS operators, and map other Observables into view models using RxJS.

You will also learn about Angular Signals and how Signals are used to reactively manage synchronous data streams. Lastly, you will learn how to combine RxJS and Signals, when to use RxJS, and when Signals reign supreme.

By the end of this chapter, you'll be able to reactively manage data streams and know everything there is to know about Angular Signals.

This chapter will cover the following topics:

- What is reactive programming?
- Reactive programming using RxJS
- Reactive programming using Angular Signals
- Combining Signals and RxJS

What is reactive programming?

Reactive programming is a declarative programming paradigm like functional, modular, procedural, or **object-orientated programming (OOP)**. A **programming paradigm** is a set of rules and principles specifying how you write your code. It is similar to architectural and design patterns, but they operate on a different level of abstraction.

Programming paradigms are high-level concepts that dictate the overall style, structure, and approach to writing code, whereas architectural and design patterns provide reusable templates or blueprints for structuring code, handling communication between components, managing relationships, and solving other common design challenges within your code base.

Reactive programming deals with data streams and the propagation of changes. In simple terms, reactive programming dictates how you handle events and data changes that can happen at any given time, also known as **asynchronous changes**. As the name already implies, you are reacting to changes with reactive programming. With reactive programming, the dependent parts of your code will automatically be notified of events and data changes so that these parts can react to changes automatically. You can think of reactive programming as a system where changes are pushed to the parts of your code that need to react upon the changes instead of you pulling the current state, checking if it changed, and then updating the dependent code accordingly.

Highlights of reactive programming

Reactive systems optimize resource allocation by subscribing to and processing data only when it changes, reducing unnecessary processing and improving overall system performance. This approach allows for a more responsive and scalable system while conserving computational resources, making it particularly advantageous for handling asynchronous data streams efficiently. Some common examples where you can use reactive programming to handle data streams and events are HTTP requests, form changes, and browser events such as clicks, mouse movements, and keyboard events.

Reactive programming also promotes composability by enabling developers to easily compose complex behaviors from simpler ones. Through operators such as `map`, `filter`, `merge`, and more, reactive systems allow you to transform, combine, and manipulate data streams.

This inherent modularity empowers developers to build more modular and flexible applications where different data streams can be seamlessly integrated, transformed, and adapted to create sophisticated and easily maintainable systems. This emphasis on composability fosters code reusability and promotes the creation of highly scalable and adaptable applications.

Another important part of reactive programming is handling events and data changes that are non-blocking; this is mainly where the performance boost comes from when you start programming reactively. **Non-blocking code** ensures multiple tasks, events, and data changes can be executed in parallel. In other words, non-blocking code runs the code without waiting for the task to finish, so your code continues to the next lines of code directly after the task has started. In contrast, **blocking code** will wait until the code has finished before moving to the next line of code.

Drawbacks of reactive programming

Reactive programming is great, but it is not only sunshine and roses; there are also some drawbacks to reactive programming.

Reactive systems can become complex, and the learning curve, especially for junior developers, can be steep. Besides some difficult concepts, reactive programming can be complicated to debug and harder to test with automated tests. Especially when you're writing unit testing for your applications, a highly reactive system can give you some headaches.

You now know what reactive programming is and how it can improve your application's performance and allow you to handle events and data streams effectively. Reactive programming provides easy composability of data streams and events and runs your code in a non-blocking approach. You also know what challenges reactive programming can bring to your code base and that it can be challenging for junior developers to understand some reactive patterns. Now, let's learn more about how reactive programming is used within Angular applications.

Reactive programming in Angular

Reactive programming is integral to the Angular framework. Angular heavily relies on *Observables* (we will explain Observables in detail in the next section) and has the **RxJS** library built into it to manage Observable data streams in a robust and composable manner. **Observables** are a reactive design pattern because you have the Observable, or publisher of data, and the subscribers or receivers of the data stream. Subscribers are automatically notified when the Observable emits a new value and can act upon it accordingly.

Observables are utilized in many different aspects within the Angular framework. Here are some examples where you can find Observables within the Angular framework:

- **HTTP requests**: Within the Angular framework, HTTP requests return Observables by default. HTTP requests in vanilla JavaScript are handled with Promises.

- **Router events**: The Angular Router class exposes an `events` Observable. With the router `events` Observable, you can listen for events such as `NavigationStart`, `NavigationEnd`, `GuardCheckStart`, and `GuardCheckEnd`.

- **Reactive forms changes**: Within reactive forms and form controls, a `valueChanges` Observable is exposed by the Angular framework. With this Observable, you can react to changes within the form or form fields.

- **ViewChildren changes**: When you use the `ViewChildren` decorator, the `QueryList` object it returns has a `changes` event. This `changes` event is an Observable that notifies you when the selection that's made by `ViewChildren` has changed.

These examples are just some instances where the Angular framework relies on Observables. In the code of applications made with Angular and within libraries that are commonly used in combination with Angular, you'll come across Observables in many instances.

Besides Observables, Angular also uses reactive programming in other ways, such as when handling browser events with the @Hostlistener() decorator:

```
@HostListener('document:keydown', ['$event'])
handleTheKeyboardEvent(event: KeyboardEvent) { ...... }
```

In the preceding code snippet, we used the @Hostlistener() decorator to listen for keydown events and handle them reactively. Another place where Angular uses the reactive programming paradigm is with **Angular Signals**, which was introduced in Angular 16. Angular Signals is a system that tracks value changes and notifies interested consumers accordingly. The Signals API comes with computed properties and effects that are automatically updated or run when a Signal value changes. Signals are excellent for handling synchronous values reactively, whereas RxJS shines in handling asynchronous data streams. We will dive deeper into the topic of Signals in the *Reactive programming with Angular Signals* section of this chapter.

Now that you know that Angular has reactive programming embedded within the framework, from Observables and RxJS to event handling and the new Angular Signals, let's move on to the next section and learn how to use RxJS to its fullest within Angular applications.

Reactive programming using RxJS

Regarding reactive programming in the context of Angular applications, RxJS stands at its core. **RxJS** is short for **Reactive Extensions Library for JavaScript**. As the name reveals, it's a library for handling reactivity within JavaScript, and it is built-in and used by default within the Angular framework.

RxJS is used to create, consume, modify, and compose asynchronous and event-based data streams. At its core, RxJS revolves around four major concepts: Observables, Observers, Subjects, and operators. Let's dive deep into these concepts individually, starting with Observables.

What are Observables?

Observables are the cornerstone of RxJS. You can see Observables as a stream or a pipeline that emits different values over time in an asynchronous manner. To receive the values emitted by the Observable data stream, you subscribe to the Observable.

Now, imagine an Observable data stream as a water pipe. When you turn on the tap (subscribe), water (data) flows through the pipe (Observable), and you receive drops of water (values) at your end. The water (data) might already be flowing before you turn on the tab (subscribe) unless you installed a special system; you will only receive water (data) from the moment you have your tab running (subscribe) up to the moment you close the tab (unsubscribe). The water (data) might keep running to other tabs (subscribers) if they are open. So, in short, to receive values from the data flow, you need to subscribe, and all values emitted before you subscribe are lost unless you have special logic in place to store these values. To stop receiving values, you need to unsubscribe and the data emitting the values flows like a stream.

Observables come in two types: *hot and cold Observables*. **Cold Observables** are unicast, meaning they start fresh for each subscriber. It's like a movie on Netflix; when someone starts the movie, the movie will start from the beginning. If someone else starts the same movie on another account or TV, the movie will also start from the beginning. Everyone watching gets their own unique viewing experience from the moment they start watching.

On the other hand, **hot Observables** are multicast, meaning there is one stream of data broadcasted to every subscriber. Hot Observables can be compared to live television. Different people can tune in to the live show (the data stream), but everything you already missed is gone and will not be replayed for you. Everyone watching experiences the same content simultaneously, even if they didn't start watching from the beginning.

Now that you clearly understand Observables and know the difference between hot and cold Observables, let's learn about Observers.

Subscribing to Observables with Observers

Observers are the entities that subscribe to an Observable and receive the Observable values in the data stream. You can think of the Observer as the person (or subscriber) watching a live show or putting on a Netflix movie. Subscribers have two tasks: subscribing to the streams they want to receive and unsubscribing from streams they do not want or need to receive anymore.

One of the most crucial parts is successfully unsubscribing from subscriptions the Observer doesn't need anymore. Not unsubscribing Observables is probably the most considerable risk when using Observables in your code and is the most common source of problems. You will run into memory leaks if you don't clean up your subscription correctly. Memory leaks will result in strange behavior and a slower application and can eventually crash your application if it runs out of memory.

Let's imagine that you have a subscription to a magazine that is delivered to your house every week. If you move to a new address, you need to *unsubscribe* from the subscription and create a *new subscription* on the new address. If you don't unsubscribe from the old address and only begin a new subscription for the new address, you will start to pay double, and the magazine will be delivered to both addresses. If you keep repeating this process, you will eventually run out of money, and your life will crash. Within your applications, it's the same, only you don't pay with money but with memory.

When you create a subscription inside a component, you must unsubscribe when the component is destroyed. Otherwise, the subscription will keep running. The next time you open the same component, a second subscription starts because the old one is still running. As a result, all values will be received by two Observers. If you keep repeating this process, you'll end up with many Observers receiving the same value while you only need a single Observer to receive the values. When there are too many Observers, the application will run out of memory to process all the values, and the app will crash.

Unsubscribing from Observables

Unsubscribing from Observables can be done in many different ways. Most commonly, in an Angular application, you can unsubscribe inside the ngOnDestroy life cycle hook. You can unsubscribe manually like this:

```
ngOnInit() {
   this.observable$.subscribe(…)
}

ngOnDestroy() {
   this.observable$.unsubscribe();
   this.observable$.complete();
}
```

In the preceding code, we subscribed to an Observable inside the ngOnInit life cycle hook and manually unsubscribed in the ngOnDestroy life cycle hook by calling the unsubscribe() and complete() methods on the Observable.

This approach may not be the best option when multiple Observables are inside your component. If you have five Observables, you must unsubscribe and complete all five Observables manually. Doing all subscriptions manually increases the risk of missing an Observable, and it will result in a large ngOnDestroy method with a lot of repetitive code inside.

You also need a local property for all Observables, further polluting your file with boilerplate code. In that case, you would need to save the Observable subscription in a property so that you can unsubscribe from the subscription in the ngOnDestroy method.

Another better option is to create a Subscription object and add all your subscriptions to this object by using the add() method on the Subscription object. In this case, you only have to unsubscribe from the Subscription object, and it will unsubscribe and complete all subscriptions that have been added to the Subscription object.

Here's an example of how you can use the Subscription object approach:

```
subscriptions = new Subscription()

ngOnInit() {
  this.subscriptions.add(this.observableA$.subscribe(…));
  this.subscriptions.add(this.observableB$.subscribe(…));
}

ngOnDestroy() {
  this.subscriptions.unsubscribe();
}
```

In the preceding code snippet, we created a subscriptions property and assigned it to the Subscription object. Next, we used the subscriptions property and added all the subscriptions to the subscriptions property using the add method of the Subscription class. Lastly, inside ngOnDestroy, we called the unsubscribe() method on the subscriptions object, which will unsubscribe and complete all the inner subscriptions.

Using the Subscription class is a good approach, but the syntax of adding the active subscriptions to the Subscription class looks messy. When you start to use RxJS pipeable operators, there is an approach that is more in line with the rest of your code. We will discuss pipeable operators in more detail later in this section, but for now, I want to show you how you can use them to unsubscribe automatically from your subscriptions.

Unsubscribing with the takeUntil() operator

First, we will take a look at the takeUntil() operator. You can add the takeUntil() operator inside the RxJS pipe method, which is changed on your Observables. The takeUntil() operator unsubscribes your subscriptions automatically upon a trigger. This trigger is commonly an RxJS Subject (we will discuss Subject in more detail later in this section). When we call the next method on this Subject, the takeUntil() operator will be triggered and unsubscribe from the subscription:

```
private destroy$ = new Subject<void>();
ngOnInit() {
  this.observable$.pipe(takeUntil(this.destroy$))
  .subscribe(……)
}
ngOnDestroy() {
  this.destroy$.next();
  this.destroy$.complete();
}
```

In the preceding example, we created a `Subject` Observable and named it `destroy$`. Next, we used the `pipe()` function on our Observable, and inside the `pipe()` function, we added the `takeUntil()` operator. The `takeUntil()` operator received the `destroy$` property as a parameter and will be triggered once we call the `next()` method on the `destroy$` property. Lastly, inside the `ngOnDestroy` life cycle method, we called the `next()` method on the `destroy$` property and finished by calling the `complete()` method to complete the `destroy$` Observable itself.

Using the `takeUntil()` operator is the preferred solution for many Angular developers to unsubscribe. This is because it works perfectly well in combination with the RxJS `pipe` function and other pipeable operators. There is one last option to unsubscribe that I want to show: the `takeUntilDestroyed()` operator.

Unsubscribing with the takeUntilDestroyed() operator

The `takeUntilDestroyed()` operator was added by Angular in version 16. It can be used to automatically unsubscribe when the component is destroyed. When you declare your subscription inside the injection context (inside the constructor or where you declare your properties), you only have to add the `takeUntilDestroyed()` operator, and it will manage everything for you:

```
data = observable$.pipe(takeUntilDestroyed()).subscribe(…);
```

When you declare the subscription in another place, such as inside `ngOnInit`, you must provide the `takeUntilDestroyed()` operator with a reference, `DestroyRef`. Here, `DestroyRef` is the `ngOnDestroy` life cycle in the form of an injectable:

```
protected readonly destroy = inject(DestroyRef);

ngOnInit() {
   this.observable$.pipe(takeUntilDestroyed(this.destroy)).
subscribe(…);
}
```

As you can see, we created a property and assigned it to `DestroyRef`. We add this property as a function parameter to the `takeUntilDestroyed()` operator. That is all you need to do, and it will unsubscribe and complete your subscriptions automatically.

You now know why it's essential to unsubscribe from Observables and how to do so using different approaches. Now, we will move on to the next major concept of RxJS: `Subject`.

Using special Observables – RxJS Subjects

The **RxJS Subject** is a special type of Observable. `Subject` Observables are **multicast Observables**, allowing you to emit values to multiple Observers simultaneously. `Subject` Observables are like `EventEmitter` Observables, which maintain a registry of all listeners.

Every `Subject` is an Observable, meaning you can subscribe to `Subject` to receive the values it emits. Each `Subject` is also an internal Observer object with the `next()`, `error()`, and `complete()` methods. The `next()` method is called to emit the next value in the data stream, `error()` is called automatically when an error occurs, and `complete()` can be called to complete the data stream.

Within RxJS, there are four different `Subjects`. Let's take a look.

Subject

This is the basic RxJS `Subject` type. The `Subject` class allows you to create a hot Observable data stream. You can emit a new value using the `next()` method; the `Subject` class has no initial value or memory of the values that have already been emitted. Subscribers will only receive values that are emitted after they are subscribed; anything emitted before that point will not be received:

```
const subject = new Subject<number>();
subject.subscribe({next: (v) => console.log(`A: ${v}`)});
subject.next(1);
subject.subscribe({next: (v) => console.log(`B: ${v}`)});
subject.next(2);

// Logs:
// A:1, A:2, B:2
```

As shown in the preceding example, we use a `Subject` class to emit two values (1 and 2). The first value is emitted after the first subscriber (A), and the second value is emitted after both subscribers have subscribed. Because of this, subscriber A receives both values, while subscriber B only receives the second value.

A good use case for the `Subject` class is when multiple Observers must respond to a specific event, such as a selection or a changing toggle. The `Subject` class can be used if components only have to react to the change if the components are active during the event. Now that you know how `Subject` works, let's examine `BehaviorSubject`.

BehaviorSubject

The `BehaviorSubject` class extends the `Subject` class and has two main differences from the regular `Subject`. The `BehaviorSubject` class receives an initial value and stores the last emitted value. When a subscriber subscribes to `BehaviorSubject`, the subscriber will immediately receive the last emitted value. When no value is emitted, the subscriber will receive the initial value instead.

The `Subject` class is good for emitting values that have to notify subscribers when an event happens, such as when multiple Observers need to react when something is added. The `BehaviorSubject` class, on the other hand, is well suited for values with state, such as `lastAddedItem`, where subscribers receive the last added item. Here, `lastAddedItem` will always emit the last item that has been added. In contrast, an `itemAdded` event using a `Subject` class will only notify subscribed Observers the moment the item is added and not after the fact:

```
const subject = new BehaviorSubject(0); //initial value 0
subject.subscribe({next: (v) => console.log(`A: ${v}`)});
subject.next(1);
subject.subscribe({next: (v) => console.log(`B: ${v}`)});
subject.next(2);

// Logs:
// A:0, A:1, B:1, A:2, B:2
```

As you can see, the `BehaviorSubject` class receives an initial value; in our case, the initial value is 0. When subscriber A subscribes to `BehaviorSubject`, the subscriber immediately gets the initial value, 0, and logs the value. After subscriber A has subscribed, we emit a new value: 1. This new value is received and logged by subscriber A.

Next, subscriber B subscribes to `BehaviorSubject`. Because 1 is the last emitted value, subscriber B gets and logs it. Lastly, we emit a new value, 2, which is received and logged by subscribers A and B.

Now that you know how `BehaviorSubject` works and how it differs from the regular `Subject`, let's learn about `ReplaySubject`.

ReplaySubject

The `ReplaySubject` class is also an extension of the regular `Subject` class and behaves a bit like `BehaviorSubject` with some differences. The `ReplaySubject` class also stores values, just like `BehaviorSubject`, but unlike `BehaviorSubject`, `ReplaySubject` can store more than one value and doesn't have an initial value.

Instead of an initial value, the `ReplaySubject` class receives a buffer size as a parameter. The buffer size determines how many values the `ReplaySubject` class stores and shares with a new subscriber upon subscription. Besides the buffer size, the `ReplaySubject` class can take a second parameter to determine how long the `ReplaySubject` class will store the emitted values in the buffer of `ReplaySubject`:

```
const subject = new ReplaySubject(100, 500);
subject.subscribe({
  next: (v) => console.log(`A: ${v}`),
});
```

```
let i = 1;
setInterval(() => subject.next(i++), 200);

setTimeout(() => {
  subject.subscribe({
    next: (v) => console.log(`B: ${v}`),
  });
}, 1000);

// Logs
// A:1, A:2, A:3, A:4, A:5, B:3, B:4, B:5, A:6, B:6
// ...
```

In the preceding example, we have a `ReplaySubject` class with a buffer of `100` and a time window of `500` milliseconds. Subscriber A subscribes before we emit the first value. Next, we create an interval that emits a new number every 200 milliseconds. As a result, subscriber A will receive and log a new value every 200 milliseconds.

Lastly, we create a timeout of 1 second and add the second subscriber. Because we have a time window of `500` milliseconds, subscriber B will immediately receive all values that are emitted after the first `500` milliseconds – that is, the timeout of 1 second that has passed minus the time window of the `ReplaySubject` class.

As a result, subscriber A logs 1 to 5; after 1 second, subscriber B joins and immediately receives values 3, 4, and 5. After subscriber B receives the replay values, both subscribers receive all values emitted after that point.

AsyncSubject

The `AsyncSubject` class is also an extension of the regular `Subject` class. The `AsyncSubject` class only emits the last value to all its subscribers and only when the Observable data stream is completed:

```
const subject = new AsyncSubject();
subject.subscribe({
 next: (v) => console.log(`A: ${v}`),
});
subject.next(1);
subject.next(2);
subject.subscribe({
  next: (v) => console.log(`B: ${v}`),
});
subject.next(3);
subject.complete();
// Logs:
// A:3, B3
```

In the preceding example, you can see that `AsyncSubject` only emitted the last value that was emitted before we completed the data stream. First, subscriber A subscribed to `AsyncSubject`. Next, we emitted two values, and then subscriber B subscribed to `AsyncSubject`. Lastly, we emitted the third value and completed the Observable stream. After we complete the stream, the last value is emitted to and logged by subscribers A and B.

Now, you know about Observables, Observers, and `Subjects`. You know that there are hot and cold Observables and the difference between the two. You also learned that Observers subscribe to Observables and how to unsubscribe from Observable data streams. You discovered that `Subjects` is a special kind of Observable that's used to multicast values and that they can emit values using the `next()` method. Lastly, you learned about the four different `Subject` types and saw how you can visualize their differences. Next, we will learn about the last major concept in RxJS: operators.

Using and creating RxJS operators

In this section, you will learn about **RxJS operators**. You will learn what operators are, what types of operators there are, and how to use some of the most commonly used operators in the RxJS library. You will also learn how to create your own RxJS operators and combine multiple operators into a single operator.

While Observables are the foundation of the RxJS library, operators are what make the library so useful and powerful for handling Observable data streams. Operators allow you to easily compose and handle complex asynchronous code declaratively.

Types of operators

RxJS operators come in two different types: creational and pipeable operators.

In short, creational operators can be used to create new Observables with a standalone function, whereas pipeable operators can be used to modify the Observable stream. Let's explore both in more detail, starting with creational operators.

Creational operators

Creational operators can be used to create a simple Observable and to create a new Observable by combining multiple existing Observables. The simplest and most straightforward example of a creational operator is the `of()` operator. The `of()` operator takes in one or more comma-separated values and turns these values into an Observable stream that emits one value after the other:

```
of(1, 2, 3).subscribe((v) => console.log(`value: ${v}`));
```

In the preceding example, we declare the of() operator with three values: 1, 2, and 3. We subscribed to the Observable stream that was created with the of() operator and logged the values. This subscription will result in three separate logs: value: 1, value: 2, and value: 3. As you can see, the of() operator is a pretty simple and straightforward way to create an Observable stream.

Another commonly used creational operator is the from() operator. The from() operator creates an Observable stream of an array, iterable, Promise, or strings. If you transform a string into an Observable stream using the from() operator, the string will be emitted character by character. Here's an example of using the from() operator:

```
from([1, 2, 3, 4, 5]).subscribe(val => console.log(val));
//output: 1,2,3,4,5

from(new Promise(resolve => resolve('Promise to Observbale!'))).
subscribe(val => console.log(val));
//output: Promise to Observbale!
```

In the preceding example, we used the from() operator to create an Observable stream from an array and a Promise.

Another useful creational operator is the fromEvent() operator. The fromEvent() operator creates an Observable from an event target such as *click* or *hover*:

```
fromEvent(document, 'click')
```

As you can see, the fromEvent() operator takes two arguments. The first is the target element – in our case, we took the document. Then, you declared the event you wanted to listen for; in our example, this is a click event.

With that, you've learned how to create a new Observable stream from scratch using creational operators. Next, you will learn how to create a new Observable stream by combining multiple existing Observable streams.

Creating an Observable from multiple Observable streams

As your Angular applications grow and the state of these applications becomes more complex, you often find yourself in a situation where you need the result of multiple Observable streams simultaneously. When you need the result from various Observable streams, you may be tempted to create nested subscriptions, but this isn't a good solution since nested subscriptions can lead to strange behavior and hard-to-debug bugs.

In scenarios where you need the result of multiple Observables, you can use creational RxJS operators that focus on combining various Observables into a new single Observable stream. When using these operators, the combined Observables are referred to as **inner Observables,** and the new combined Observable is called the **outer Observable**. Let's examine some of the most commonly used operators that create a new Observable based on multiple other Observables, starting with the `combineLatest()` operator.

The `combineLatest()` operator is best used when you have multiple long-lived Observables and need the values of all these Observables to construct the object or perform the logic you want. The `combineLatest()` operator will only output its first value when all of its inner Observables output at least one value; after that, `combineLatest()` outputs another value each time one of the inner Observables emits a new value. The `combineLatest()` operator will always use the last emitted value of all its inner Observables:

```
const amountExclVat = of(100);
const vatPercentage = of(20);

combineLatest([amountExclVat, vatPercentage]).subscribe({
  next: ([amount, percentage]) => {
    console.log(<Total:', amount * (percentage / 100 + 1));
  }
});
```

As you can see, we provided `combineLatest()` with an array containing two Observables: one Observable with the amount excluding VAT and another Observable containing the VAT percentage. We need both Observable values to log the amount, including VAT. Instead of creating a nested subscription, we handled this with `combineLatest()`.

Inside the `combineLatest()` subscription, we also declare an array for the value of the Observable stream. We used `amount` and `percentage` as values, but you can name these properties however you like. Alternatively, you can use a different syntax and provide an object to `combineLatest()` instead of an array:

```
combineLatest({ amount: amountExclVat, percentage: vatPercentage
}).subscribe({
  next: (data) => { console.log('Total:', data.amount * (data.
percentage / 100 + 1)) }
});
```

Now, let's consider another example where we emit different values over time so that you get a better understanding of how `combineLatest()` works and when and what values it will emit:

```
const a = new Subject();
const b = new Subject();
combineLatest([a, b]).subscribe({
```

```
  next: ([a, b]) => { console.log(<data>, a, b) }
});

a.next(1);
setTimeout(() => { b.next(2) }, 5000);
setTimeout(() => { a.next(10) }, 10000);
```

In the preceding example, it takes 5 seconds before the combineLatest() operator emits the first value; this is because, after 5 seconds, both Observable a and b have emitted a value. Even though Observable A directly emits a value, combineLatest() will only emit a value after both A and B have emitted at least one value.

After 5 seconds have passed and both Observables have emitted a value, combineLatest() will emit a value, and we log data: 1, 2. After both Observables emit a value, combineLatest() will emit a new value whenever one of its Observables emits a new value. So, when Observable A emits a new value after another 5 seconds have passed, we log data: 10, 2 inside the subscription of combineLatest().

If, for example, you first emitted two values with Observable A (1, 10) and then emitted a value with Observable B (2), combineLatest() will only emit one value, data: 10, 2. This is the case because both A and B need to emit a value before combineLatest() starts emitting values.

Now that you have a good idea of how combineLatest() works and how to use RxJS to create a new Observable based on multiple Observables, let's explore other operators that create an Observable from multiple Observables:

- **forkJoin**: The forkJoin() operator is best used when you have multiple Observables and are only interested in the final value of each of these Observables. This means that each Observable has to be completed before forkJoin() emits a value. A good example of when forkJoin() is useful is when you must make multiple HTTP requests and only want to do something when all requests return a result. The forkJoin() operator can be compared with Promise.all(). It's important to note that if one or more of the inner Observables has an error (and you don't catch that error correctly), forkJoin() will not emit a value:

  ```
  forkJoin({ posts: this.http.get('…'), post2: this.http.
  get('…') }).subscribe(console.log);
  // Logs: { posts: …… , post2: …… }
  ```

- **concat**: The concat() operator is used when you have multiple inner Observables, and the order of the emission and completion of these inner Observables is essential. So, if you have two inner Observables, concat() will emit all the values of the first inner Observable until that Observable is completed. All the values that have been emitted by the second Observable before the first Observable has been completed will not be emitted by concat().

When the first Observable has been completed, the `concat()` operator subscribes to the second Observable and starts to emit the values emitted by the second Observable. If you have even more inner Observables, this process will be repeated, and `concat()` will subscribe to the next Observable when the previous Observable is completed:

```
const a = new Subject();
const b = new Subject();
concat(a, b).subscribe(console.log);

a.next(1);
a.next(2);
b.next(3);
a.complete();

b.next(4);
b.complete();
// Logs: 1, 2, 4
```

- **merge**: The `merge()` operator combines all inner Observables and emits the values as they come in. The `merge()` operator doesn't wait for all Observables to emit a value, nor does it care about the order. When one of the Observables emits a value, the `merge()` operator will process it:

```
const a = new Subject();
const b = new Subject();
merge(a, b).subscribe(console.log);
b.next('B:1');
a.next('A:1');
a.next('A:2');
b.next('B:2');
// Logs: B:1, A:1, A:2, B:2
```

You now know how to create Observables with creational operators. You know there are creational operators such as `of()` and `from()` to create simple Observables and creational operators such as `combineLatest()` to create a new Observable based on multiple inner Observables.

Next, we will learn about pipeable operators and how they can be used to filter, modify, and transform Observable streams.

Pipeable operators

Pipeable operators take in an Observable as input and return a new and modified Observable without modifying the original Observable. When you subscribe or unsubscribe to the piped Observable, you also subscribe or unsubscribe to the original Observable. Pipeable operators can filter, map, transform, flatten, or modify the Observable stream. For example, pipeable operators can be used to unsubscribe upon a trigger automatically, take the first or last emission of an Observable steam, only emit an Observable value if specific conditions are met, or map the output of the Observable stream into a new object.

Using pipeable operators starts with using the `pipe()` function on an Observable. The `pipe()` function acts like a path for your Observable data, guiding it through different tools called operators. It's like how materials in a factory move through various stations before becoming a finished product. Here, your data can go through these operators, where you can change it, pick out specific parts, or make it fit your needs. It's a common scenario that developers use four, five, or even more operators inside a single `pipe()` function.

Let's examine an example and learn about some commonly used pipeable operators:

```
const observable = of(1, 1, 2, 3, 4, 4, 5);

observable.pipe(
  distinctUntilChanged(),
  filter(value => value < 5),
  map(value => value as number * 10)
).subscribe(results => {console.log('results:', results)});
// Logs: 10, 20, 30, 40
```

In the preceding code, we created an Observable using the `of()` operator. On the Observable, we use the `pipe()` function with three different pipeable operators declared inside the pipe function: `distinctUntilChanged()`, `filter()`, and `map()`. At the end of the pipe function, we subscribe to the Observable stream. The values of the Observable stream move through the pipe and perform the operators on them one by one before ending up in the subscribe block of our code.

The first `distinctUntilChanged()` operator checks if the Observable value differs from the previous and filters it out if the value is the same as the last emitted value. Next, the `filter()` operator works similarly to the filter function on an array; in our case, we filter out all values that aren't smaller than 5. Lastly, we use the `map()` operator; this is also similar to the map function on an array and lets you map the value to a new value; in our case, we multiply by a factor of 10. After applying all our operators, the Observable values that are logged in the subscription are 10, 20, 30, and 40; all other values of our Observable are filtered out.

As you can see, the pipeable operators are performed one after another, guiding the Observable value through a pipe where changes are applied to the value until it reaches the subscription or is filtered out. The `distinctUntilChanged()`, `filter()`, and `map()` operators are some of the most commonly used operators. In this chapter, you also learned about the `takeUntil()` and `takeUntilDestroyed()` operators, which are also commonly used.

Now, let's continue by exploring some other powerful and commonly used operators and scenarios when pipeable operators are helpful, starting with flattening operators.

Flattening multiple Observable streams using flattening operators

As we've seen earlier in this section, sometimes, you need the value of multiple Observables. In some cases, you need all these values at once; in these scenarios, you can use the creational operators that create a new Observable based on multiple inner Observables.

But in other scenarios, you first need the value of one Observable to pass as an argument to another Observable. These scenarios where you have an outer Observable and an inner Observable, where the inner Observable relies on the value of the outer Observable, are commonly referred to as **higher-order Observables**. In most cases, the correct way to handle higher-order Observables is to flatten them into a single Observable stream. You can achieve this using **flattening operators** such as `concatAll()`, `mergeAll()`, `swtichAll()`, and `exhaustAll()`. To get a better understanding of this concept, let's look at some examples.

Let's say you have an Observable yielding an API URL. Next, you want to use this URL to make an HTTP request. In actuality, you're only interested in the result of the API request and not so much in the result of the Observable yielding the URL. The URL is only needed to make the API request, and the API response is required to render your page or perform some logic. One approach would be to nest the two subscriptions, but as you've learned, this isn't a good approach. The correct solution is to use a flattening operator to flatten the Observable stream into a single stream:

```
ngOnInit() {
  this.urlObservable.pipe(
    map((url) => this.http.get(url)),
    concatAll()
  ).subscribe((data) => { console.log('data ==>', data) })
}
```

As you can see, we have an outer Observable receiving an API URL and using two pipeable operators on this Observable. First, we use the `map()` operator to take the result of the URL Observable and use it to make the API request, which results in our second Observable. Next, we use the `concatAll()` operator to flatten the two Observables into a single Observable, only returning the result of the API request.

Inside the subscription, we log the result, which will be the data that's returned by the API call. You can simplify this code even more by using the combined operator, `concatMap()`, which combines the `map()` and `concatAll()` operators into a single operator:

```
this.urlObservable.pipe(
    concatMap((url) => this.http.get(url)),
).subscribe(……)
```

These combined operators exist for all four flattening operators, so you have the following operators:

- `concatMap()`
- `mergeMap()`
- `switchMap()`
- `exhaustMap()`

Now that you've seen how you can use a flattening operator and that there are map operators that combine the map and flattening operators, let's learn about the difference between them.

The concatMap() operator

The `concatAll()` operator is used when you want the first value of the outer Observable and all its inner concatenated Observables to complete before the second value of the outer Observable and its inner Observables are processed. Let's consider the following example:

```
const clicks = fromEvent(document, 'click');
clicks.pipe(
    concatMap(() => interval(1000).pipe(take(4))),
).subscribe(number => console.log(number));
```

In the preceding code, we use the `fromEvent()` creational operator to create an Observable whenever we click the browser document (that would be any place in our app). Next, we use `concatMap()` to map the result of the click Observable into a new Observable using the RxJS `interval()` creational operator. The `interval()` operator will emit sequential numbers starting at zero; in our case, it will emit the following number every 1000 milliseconds.

We also used the `take()` pipeable operator on the interval Observable. This limits the number of emissions we take to 4, so the interval Observable will emit 0, 1, 2, and 3 as values and be unsubscribed and completed by the `take()` operator afterward.

Because we use the `concatMap()` flattening operator, when we click twice on the screen, both the outer and inner Observables will be triggered two times, but the first click Observable and its inner Observables will be processed first and only when that is completed the second sequence will start. So, our subscription part will log 0, 1, 2, 3, 0, 1, 2, 3.

The mergeMap() operator

Now, let's consider the same scenario with the other flattening operators, starting with the mergeMap() operator:

```
const clicks = fromEvent(document, 'click');
clicks.pipe(
  mergeMap(() => interval(1000).pipe(take(4))),
).subscribe(x => console.log(x));
```

In the preceding example, we only changed concatMap() for the mergeMap() operator, yet the result will be completely different. The mergeMap() operator will not wait for the first inner and outer Observables to complete but will process the values as they come in.

So, if you click on the screen, wait for 2 seconds, and then click on the screen again, the values of the second click and its inner interval Observable will start to come in before the first stream has completed. If you click the first time on the screen, the first log will come in after one second and another one for every second.

Then, when you click again after a second, the first log of the second stream will come in and log another value for every second after that. In this case, the result of all logs would be 0, 1, 2, 0, 3, 1, 2, 3. As you can see, the result is entirely different from the result we had with concatMap(). Now, let's see what happens when we change mergeMap() to switchMap().

The switchMap() operator

The switchMap() operator will switch the Observable stream from the first to the second stream when the second stream starts to emit values. The switchMap() operator will unsubscribe and complete the first stream so that the first stream will stop emitting values; the next stream will keep emitting until that stream is completed or until another stream comes:

```
const clicks = fromEvent(document, 'click');
clicks.pipe(
  switchMap(() => interval(1000).pipe(take(4))),
).subscribe(x => console.log(x));
```

So, with the preceding code, if we click on the screen now, wait for 2 seconds, and then click another time, our log will look like 0, 1, 0, 1, 2, 3. As you can see in the logs, the first stream is completed the moment the second stream starts to emit values. Lastly, we have the exhaustMap() operator.

The exhaustMap() operator

The `exhaustMap()` operator will start to emit the values of the first Observable stream as soon as it starts to emit values. The `exhaustMap()` operator will not process any other Observable streams that come in while the first stream is still running. Only when the stream has been completed will new values be processed, so if you click while the first stream is still running, it will never be processed:

```
const clicks = fromEvent(document, 'click');
clicks.pipe(
  exhaustMap(() => interval(1000).pipe(take(4))),
).subscribe(x => console.log(x));
```

In the preceding example, where we used the `exhaustMap()` operator and clicked and waited for 2 seconds before we made another click, only the first click will be processed because the first stream takes 4 seconds to complete. So, when we make the second click, the first stream is not completed yet, so `exhaustMap()` doesn't process the second stream. The log of the preceding code will look like `0, 1, 3, 4`.

Lastly, it's important to note that when you use the `take()`, `takeUntil()`, or `takeUntilDestroyed()` operators inside the pipe of the outer Observable and also use flattening operators in the same `pipe()` function, the `take()`, `takeUntil()`, and `takeUntilDestroyed()` operators need to be declared after the flattening operators. The flattening operators will create their own Observables, and if you declare `take()`, `takeUntil()`, or `takeUntilDestroyed()` before the flattening operator, the Observables created by the flattening operators will not be unsubscribed and closed by `take()`, `takeUntil()`, or `takeUntilDestroyed()`.

You now know what higher-order Observables are and how you can handle them using flattening operators. You learned about combined operators that combine the `map()` and flattening operators, and you learned about using `take()`, `takeUntil()`, or `takeUntilDestroyed()` in combination with the flattening operators. Lastly, you learned about the `interval()` and `take()` operators.

Now, let's start exploring other useful pipeable operators that serve a few more straightforward use cases and scenarios.

Powerful and useful RxJS operators

You have already learned much about RxJS and seen how you can handle some complex Observable scenarios by combining or flattening Observables. You've also seen how to unsubscribe Observables or filter values using pipeable operators. We will now walk through some commonly used pipeable operators that are useful in more straightforward scenarios:

- `debounceTime()`: The `debounceTime()` operator takes a pause and waits for another value to come in within the defined timeframe. A good real-world example of this is a search or filter input field. Instead of bombarding your system with an update for every keystroke, a more efficient solution would be waiting until the user stops typing for a specific interval. This waiting can be done by using `debounceTime()`. You provide `debounceTime()` with a parameter indicating the milliseconds it should wait (`debounceTime(300)`) before processing the value; only when no new value is received within the specified timeframe will the value be passed on to the next operator or the subscription block.

- `Skip()`: The `skip()` operator can skip a fixed number of emissions. Let's say you have `ReplaySubject`, and for one of your subscriptions on `ReplaySubject`, you aren't interested in the replayed emissions, only in the new emission. In this scenario, you can use the `skip()` operator and define the number of emissions you want to skip inside the operator: `skip(5)`.

- `skipUntil()`: The `skipUntil()` operator works a bit like the `takeUntil()` operator; only it will skip the emissions until the inner Observable of `skipUntil()` receives a value. You could provide `skipUntil()` with a `Subject` class or something like an RxJS timer so that you only take values after a predefined interval has passed: `skipUntil(timer(5000))`.

- `find()`: The `find()` operator works similarly to the `find` method on an array. It will only emit the first value it finds that matches the condition you provide the `find()` operator with. So, `find((item: any) => item.size === 'large')` will only pass on the first item through the pipe where the size property is equal to `large`.

- `scan()`: The `scan()` operator is comparable to the `reduce` function on an array. It gives you access to the previous and current value and allows you to emit a new value based on the previous and current value. For example, you can combine the results or take the lowest or highest result of the two: `scan((prev, curr) => `${prev} ${curr}`, '')`. Here, we combined the previous and current values using the `scan()` operator.

With that, we have covered some of the more commonly used operators and learned how to filter, map, limit, or transform the value stream with pipeable operators. You can find them in the official documentation if you want to learn more about operators and check out a complete list: `https://rxjs.dev/guide/operators`.

Before we move on to the next section and start to learn about Angular Signals, let's finish this section on RxJS by creating combined and reusable operators.

Creating combined and reusable RxJS operators

Creating a reusable operator or combining multiple operators can easily be done by creating a function that returns an RxJS `pipe()` function. Let's say you find yourself making a filter pipe that filters odd numbers multiple times. Creating a function that does this would be easier so that you don't have to repeat the logic numerous times. You can do this by creating a function that returns an RxJS pipe implementing the filter operator with the filter logic predefined in the operator:

```
export const discardOdd = () => pipe(
   filter((v: number) => !(v % 2)),
);
```

You can now use this pipeable operator like any other operator:

```
of(1, 2, 4, 5, 6, 7).pipe(discardOddDoubleEven()).subscribe(console.
log);
```

If you want to combine multiple operators, it works the same way: you must create a function that returns an RxJS pipe and declares all the operators you want to use inside the pipe function:

```
const discardOddDoubleEven = () => pipe(
   filter((v: number) => !(v % 2)),
   map((v: number) => v * 2)
);
```

With that, you've learned all about operators and how to use pipeable and creational operators. You know how to combine and flatten multiple Observable streams and create reusable and combined operators using the `pipe()` function. You've seen how to create Observable streams and handle code reactively and asynchronously with Observables and RxJS.

Since the introduction of Angular Signals, you can also handle reactivity more synchronously, allowing you to do almost everything in your Angular applications in a reactive manner, both with synchronous and asynchronous code.

In the next section, you will learn everything about Angular Signals. You will learn what Signals are and how and when to use them.

Reactive programming using Angular Signals

We briefly discussed **Angular Signals** in *Chapter 2*, but let's reiterate that and dive a bit deeper so that you can get a good grasp of Angular Signals and how they can help you handle code more reactively.

Angular Signals was introduced in Angular 16, and it's one of the most significant changes for the framework since it went from AngularJS to Angular. With Signals, the Angular framework now has a reactive primitive in the Angular framework that allows you to declare, compute, mutate, and consume synchronous values reactively. A **reactive primitive** is an immutable value that alerts consumers when the primitive is set with a new value. Because all consumers are notified, the consumers can automatically track and react to changes in this reactive primitive.

Because Signals are reactive primitives, the Angular framework can better detect changes and optimize rendering, resulting in better performance. Signals are the first step to an Angular version with fully fine-grained and local change detection that doesn't need Zone.js to detect changes based on browser events.

At the time of writing, Angular assumes that any triggered browser event handler can change any data bound to an HTML template. Because of that, each time a browser event is triggered, Angular checks the entire component tree for changes because it can't detect changes in a fine-grained manner. This is a significant drain on resources and impacts performance negatively.

Because Signals notify interested parties of changes, Angular doesn't have to check the entire component tree and can perform change detection more efficiently. While we aren't at a stage yet where Angular can perform fully local change detection and only update components or properties with changes, by using Signals combined with OnPush change detection, you reduce the number of components Angular has to check for changes. Eventually, Signals will allow the framework to perform local change detection, where the framework only has to check and update components and properties that have changed values.

Besides change detection, Signals bring more advantages. Signal allows for a more reactive approach within your Angular code. While RxJS already does a fantastic job facilitating reactive programming within your Angular applications, RxJS focuses on handling asynchronous Observable data streams and isn't suited to handle synchronous code.

On the other hand, Signals shine where RxJS falls short; Signals reactively handle synchronous code by automatically notifying all consumers when the synchronous value changes. All dependent code can then react and update accordingly when the Signal pushes a new value. Especially when you start to utilize Signal effects and computed Signals, you can take your reactivity to the next level! Signal effects and computed Signals will automatically compute new values or run side effects when the Signal value changes, making it easy to automatically update and run logic as a reaction to the changed value of synchronous code.

Another problem that Signals solves is the infamous and dreaded `ExpressionChanged AfterItHasBeenCheckedError` error. If you've worked with Angular, changes are pretty significant you've seen this error before. This error occurs because of how Angular currently detects changes. Because the change detection on Signals is different, as Angular knows when they change and doesn't have to check for changes, the dreaded `ExpressionChangedAfterItHasBeenCheckedError` error will not occur when working with Signal values.

Signals wrap around values such as strings, numbers, arrays, and objects. The Signal then exposes the value through a getter, which allows the Angular framework to track who is consuming the Signal and notify the consumers when the value changes. Signals can wrap around simple values or complex data structures such as objects or arrays with nested structures. Signals can be read-only and writable. As you might expect, writeable Signals can be modified, whereas read-only Signals can only be read.

Now that you understand the theory behind Signals, let's dive into some examples and learn how and when to use Angular Signals within Angular applications.

Using Signals, computed Signals, and Signals effects

The best way to better understand something is to use it. So, without further ado, let's start learning about Signals by writing some code. Start by cleaning up your `expenses-overview` component, clear the entire HTML template, and remove any logic you still have in the component class. Your component class and the corresponding HTML template should be empty when you're done.

To explain Signals step by step, we will initially use hardcoded expenses inside the `expenses-overview` component. We'll start by creating an `expenses` Signals with an initial value containing an array with some expenses inside:

```
expenses = signal<ExpenseModel[]>([ ...... ]);
```

As you can see, we created a property and assigned it a `signal()` function. This function receives a parameter that sets the initial value of the Signal. In our example, we have added an array with some expenses for the initial value (you can create the mocked expenses based on `ExpenseModel`). You can manually add a type for your Signal using the arrow syntax, `<ExpenseModel[]>`, but the Signal also infers the type from the initial value. Let's use this Signal inside our HTML template to output the expenses.

You can access a Signal like any other function – you use the property name of the Signal and add function brackets after it. In general, I don't recommend using functions inside your HTML template, but Signals are an exception as they are non-computational functions; they return a value without computing anything. So, let's output our Signal inside the HTML template:

```
<div class="container">
  <h1>Expenses Overview</h1>
  <table>
    <tr>
      <th>Description</th>
      ......
    </tr>
    @for (expense of expenses(); track expense.id) {
    <tr>
      <td>{{ expense.description }}</td>
      ......
```

```
    </tr>
    }
  </table>
</div>
```

Here, we've created an HTML table and used the control flow syntax to output a table row for each expense within our `expenses` Signal. We accessed the expenses by calling our Signal with `expenses()`. You can compose your own table headers and data rows or copy the HTML and CSS from this book's GitHub repository. Now that you know how to create and use Signal values, next, you will learn how to update your Signals.

Updating Signals

A Signal can be updated by using the `set()` or `update()` method on it. The `set()` method sets an entirely new value, whereas the `update()` method allows you to use the current Signal value and construct a new value based on the current value of the Signal.

To demonstrate this, let's add a modal with `AddExpenseComponent` inside the modal. Before you add the form inside the template, let's update `AddExpenseComponent` so that it uses our new `ExpenseModel` instead of the `AddExpenseReactive` model we used in *Chapter 4*. Replace all instances of `AddExpenseReactive` with `ExpenseModel`. Now, change the `date` and `tags` fields in the `addExpenseForm` property to this:

```
date: new FormControl<string | null>(null, [Validators.required]),
tags: new FormArray<FormControl<string | null>>([
  new FormControl(''),
])
```

Now that we've updated the form so that it uses `ExpenseModel`, let's import `AddExpenseComponent` and `ModalComponent` into `ExpensesOverviewComponent` so that we can use them inside the HTML template. Next, create a new Signal in `ExpensesOverviewComponent` to control the state of the modal component:

```
showAddExpenseModal = signal(false);
```

After adding the Signal to control the modal state, you can add both the modal and expense components to the HTML template, like this:

```
<bt-libs-modal [shown]=" showAddExpenseModal()" (shownChange)="
showAddExpenseModal.set(false)" [title]="'Add expenses'">
  <bt-libs-ui-add-expense-form #form
  (addExpense)="onAddExpense($event)" />
</bt-libs-modal>
```

As you can see, when the modal outputs the `shownChange` event, we use the `set()` method on the `showAddExpenseModal` Signal to set a new value for the Signal. In this scenario, we don't care about the previous signal value because we know we want to close the modal when this event is fired. Because we don't care about the previous value, we can use the `set()` method on the Signal to set a new value. Inside the component class, we need to add the `onAddExpense` method so that we can the expense we submit in the add expense form:

```
onAddExpense(expenseToAdd: ExpenseModel) {
  this.expenses.update(expenses => [...expenses, expenseToAdd]);
  this. showAddExpenseModal.set(false);
}
```

In the preceding code, we used the `update()` method to change the `expenses` Signal and the `set()` method for the `showAddExpenseModal` Signal. We use the `update()` method for the `expenses` Signal to access the current state of `expenses` and add the new expense to the existing expenses. When we submit the form, we also want to close the modal. For this, we can use the `set()` method because we just wish to change the Signal to a `false` value and are not interested in the current value of the Signal. Lastly, we need a button to open the modal:

```
<button (click)="showAddExpenseModal.set(true)">Add expense</button>
```

After adding the button, you can open the modal and create a new expense using `addExpenseForm`.

Lastly, it's good to know that when you update a Signal using `set()` or `update()` in a component with `OnPush` change detection, the component will automatically be marked by Angular to be checked for changes. As a result, Angular will automatically update the component on the next change detection cycle.

Now that you know how to create and update Signals, let's learn about computed Signals.

Computed Signals

Computed Signals are one of the concepts that make Signals so powerful. A computed Signal is a read-only Signal that derives its value from other Signals. Because computed Signals are read-only, you can't use the `set()` or `update()` method on them. Instead, computed Signals automatically update when one or more Signals they derive their value from changes.

Let's start with a basic example to better understand computed Signals and how they work. You don't have to add this example inside `ExpensesOverviewComponent`; it's just for demonstration purposes:

```
const count: WritableSignal<number> = signal(0);
const double: Signal<number> = computed(() => count() * 2);
```

In the preceding example, we have a `count` Signal and a computed Signal named `double`. The computed Signal uses the `computed` function, and the `count` Signal is used inside the callback of the `computed` function. When the value of the `count` Signal changes to `1`, the value of the computed Signal is automatically updated to `2`.

It's important to know that the computed Signal will only update when the Signal it depends on has a new stable value. Notice how I say *stable value*; this is because Signals provide their updates asynchronously and work a bit like the RxJS `switchMap` operator, which cancels the previous data stream if a new stream comes in before the old stream is finished. So, if you update the Signal multiple times in a row without pause in between, the Signal will not stabilize its value, and as a result, the computed Signal will only run on the last value change of the Signal.

Computed signals are very powerful and also very efficient. The computed Signal will not compute a value until the computed value is read for the first time. Next, the computed Signal will cache its value, and when you read the computed Signal again, it will simply return the cached value without running any computations. If the Signal value that the computed signal uses to derive its value from changes, the computed Signal will run a new computation and update its value.

Because computed Signals cache their results, you can safely use computationally expensive operations such as filtering and mapping arrays inside the callback function of your computed Signals. Just like regular Signals, computational Signals notify all consumers when their values change. As a result, all consumers of computed Signals will show the latest computed value.

Now, let's add a computed Signal to `ExpensesOverviewComponent` for the total amount, including VAT:

```
totalInclVat = computed((() => this.expenses().reduce((total, { amount:
{ amountExclVat, vatPercentage } }) => amountExclVat / 100 * (100 +
vatPercentage) + total, 0));
```

As you can see, we used the `computed` function, and inside the callback of the `computed` function, we used the `expenses` Signal to retrieve the current list of expenses. We can use the `Array.reduce` function on the `expenses` array to retrieve the total costs, including VAT. You can access the computed signal the same as you would a regular signal:

```
this.totalInclVat()
```

Let's create a new table row in the HTML template to display the total value. You can add the table row below the `for` loop in the HTML template:

```
<tr class="summary">
  <td>Total: {{totalInclVat()}}</td>
</tr>
```

Suppose you add a new expense using the add expense form. In that case, you'll notice that the total amount is automatically updated because the computed Signal uses the `expenses` Signal to evaluate the total amount. When the `expenses` Signal changes, the computed Signal will also be updated based on the `expenses` Signal.

One last thing that is good to know about computed Signals is that only Signals that are used during the computation are tracked. For example, let's say you add a Signal to control if you show or hide the table row summary and another for the corresponding button text:

```
showSummary = signal(false);
summaryBtnText = computed(() => this.showSummary() ? 'Hide summary' :
'Show summary');
```

You can now use the following `showSummary` Signal inside the computed Signal like this:

```
totalInclVat = computed(() => this.showSummary() ? this.expenses().
reduce(
    (total, { amount: { amountExclVat, vatPercentage } }) =>
amountExclVat / 100 * (100 + vatPercentage) + total,
    0
  ) : null);
```

In this scenario, the computed Signal will only track the `expenses` Signal if the `showSummary` Signal is set to `true`. If the `showSummary` Signal is set to `false`, the expenses Signal is never reached within the `computed` function, and because of that, it will not be tracked for changes. So, if you update the `expenses` Signal while the `showSummary` Signal is set to `false`, the computed Signal will not calculate a new value.

Now that you know what computed Signals are, how to use them within your code, and how computed Signals update their values, let's explore Signal effects.

Signal effects

Signal effects are side effects that run each time a Signal changes. You can perform any logic you want inside a Signal effect. Some use cases for a Signal effect would be logging, updating local storage, showing notifications, or performing DOM manipulations that can't be handled from within the HTML template.

You can create a Signal effect by using the `effect` function and providing the `effect` function with a callback:

```
effect(() => { console.log(`Count is: ${count()}`) } );
```

Signal effects always run when the `effect` function is initialized. Furthermore, when you use a Signal inside the callback of the `effect` function, the `effect` function becomes dependent on that Signal, and the `effect` function will run each time one of the Signals it depends on has a new stable value.

It is also good to know that just as with computed Signals, a Signal effect only runs if the Signal within the `effect` function can be reached:

```
effect(() => {
  if (this.showSummary()) {
    console.log(<Updated expenses:>, this.expenses());
  }
});
```

In the preceding example, the `effect` function will not run if the `expenses` signal updates while the `showSummary` signal is evaluated to be `false`. Besides unreached Signals, you can also prevent the Signal's `effect` function from reacting to a Signal by wrapping that Signal in the `untracked` function:

```
effect = effect(() => {
  console.log(Summary:', this. showSummary ());
  console.log('Expenses:', untracked(this.expenses()));
});
```

Another good thing to know about Signal effects is that they need access to the injection context. This means you need to declare the Signal inside the constructor or directly assign it to a property where you declare your component properties. An error will be thrown when you create a Signal effect outside the injection context. If you need to declare a Signal effect outside the injection context, you can provide the effect with the injection context like so:

```
injector = inject(Injector);
effect(() => {
  console.log(<Updated expenses:>, this.expenses());
}, { injector: this.injector });
```

By default, effects clean up when the injection context where the effect is declared is destroyed. If you don't want this to happen and you need manual control over the destruction of the signal effect, you can configure the effect so that it uses manual cleanup:

```
expenseEffect = effect(() => {
  console.log(<Updated expenses:>, this.expenses());
}, { manualCleanup: true });
```

After setting `manualCleanup` to `true`, you have to call the `destroy()` function on the effect:

```
expenseEffect.destroy();
```

You can also hook into the cleanup of a Signal effect by using a callback function. This can be useful when you want to perform some logic when the Signal effect is cleaned up. Here's an example of the onCleanup callback:

```
effect((onCleanup) => {
  onCleanup(() => { console.log('Cleanup logic')})
})
```

Besides manualCleanup and the onCleanup callback, there is one last configuration option for the Signal effects. By default, you are not allowed to update Signals inside a Signal effect; this is because this can easily result in infinite executions of the Signal effect. However, you can circumvent this by setting the allowSignalWrites property on the Signal effect:

```
expenseEffect = effect(() => {
  console.log(<Updated expenses:>, this.expenses());
}, { allowSignalWrites: true });
```

Now that you know all about Signal effects, how to use them, and how you can trigger or configure them, let's learn about Signal component inputs.

Signal component inputs

Since Angular 17.1, you can also use Signals as component input instead of using the @Input () decorator. Using the **Signal input** makes your component more readable and helps you eliminate the ngOnChanges life cycle hook.

Let's see an example of a Signal input and compare it to the @Input () decorator (you don't have to add the example in the monorepo, it's just for explanatory purposes):

```
@Input() data!: DataModel; // The old way of doing things
data = input<DataModel>(); // The signal input
```

As you can see, the Signal input has a much more straightforward approach to declaring input properties. You declare a property and assign it with an input () function; optionally, you can add the arrow syntax to add a type to the Signal input. If you want to provide an initial value to your Signal input, you can provide it as a function parameter, like this:

```
data = input({ values: [......], id: 1 });
```

Just as with the input decorator, you can also make the input required, use an input alias, or create a transform function on the input:

```
data = input.required<DataModel>();
data = input({ values: [......], id: 1 }, { alias: 'product'});
data = input({ values: [......], id: 1 }, transform: sort<DataModel>);
export function sort<T>(data: T[]): T[] {
  return data.sort((a, b) => a.id - b.id)
}
```

As you can see, you can make the input required and provide the Signal input with a configuration object to give it an alias and transform function. The Signal input uses a simpler syntax, helps improve the change detection mechanism, and allows you to remove the ngOnChanges life cycle hook.

When you combine the Signal input with computed Signals, you can remove the ngOnChanges life cycle hook because all properties that need to be updated upon the input of a specific property can now automatically be updated by using a computed Signal based on the input Signal. Any additional logic you want to run can be declared inside a Signal effect that reacts to the Signal input.

Now that you know more about Signal inputs, let's learn about Signal queries, which are used to interact with HTML elements reactively.

Signal queries

Often, you need to select HTML elements from your template and interact with them inside your component class. In Angular, this is commonly achieved by using the @ContentChild, @ContentChildren, @ViewChild, or @ViewChildren decorator. In Angular 17.2, a new Signal-based approach was introduced that allows you to interact with HTML elements reactively and combine them with computed Signals and Signal effects. Using the Signal-based approach instead of decorators offers some additional advantages:

- You can make the query results more predictable.

- All Signal-based queries return a Signal and when there are multiple values, the Signals returns a regular array. The decorators return a variety of return types and when your query returns multiple values, they return a query list instead of a regular array.

- Signal-based queries can be used for HTML elements that change over time because they are rendered conditionally or outputted in a for loop. The directive approach has issues with both scenarios and doesn't automatically notify you when the template is updated.

- TypeScript can automatically infer the type of the queried HTML element or component.

Now that you know about the advantages of query-based Signals compared to the directive approach, let's explore the syntax of query-based Signals. Instead of defining a directive, the Signal-based approach works with simple functions. There are four different functions: `viewChild()`, `contentChild()`, `viewChildren()`, and `contentChildren()`. You provide the functions with your query selector, similar to the query selectors you can use in combination with decorators. Here's an example of how to use Signal queries:

```
@Component({
  template: `
      <div #el></div>
      <my-component />
    `
})
export class TestComponent {
  divEl = viewChild<ElementRef>('el');
  cmp = viewChild(MyComponent);
}
```

In the preceding example, we used the `viewChild()` query Signal to retrieve the `<div>` element with the `#el` ID defined on it and the `<my-component>` element. Alternatively, if you want to retrieve multiple elements with the same selector, you can use the `viewChildren()` function.

With that, you know how to query template elements using the new Signal-based approach. You also learned what advantages the new Signals-based approach has over decorators. Decorators can still be used if you prefer them or have them within your codebase.

In this section, you learned about Signals. You learned how to create, read, and update Signals using the `set()` and `update()` methods. Then, we added some Signals in `ExpensesOverviewComponent` and learned about computed Signals. Lastly, you learned about Signal effects, Signal component inputs, and query Signals.

In the previous section, you learned about RxJS; in the next section, you will learn how you can combine RxJS and Signals.

Combining Signals and RxJS

In this chapter, you've seen how Signals and RxJS can help you manage data changes reactively. Both Signals and RxJS allow you to react when values change and create new values by combining multiple data streams or performing side effects based on data changes. So, the following questions might arise: Do signals replace RxJS? And when do I use Signals, and when do I use RxJS?

RxJS can sometimes feel daunting and complex, so some developers might be tempted to replace RxJS with Signals completely. While this might be possible for some applications, both RxJS and Signals have their place within your applications and solve different problems and needs. In many scenarios, you can devise a solution for your problem using Signals or RxJS, but one of the two will do a better job at solving the problem and will handle it with much fewer lines of code. Signals are not here to replace RxJS, yet Signals will complement it and, in many scenarios, work together with your RxJS code.

Because Signals and RxJS are supposed to co-exist, Angular created two RxJS interoperability functions: **toSignal** and **toObservable**. While these functions are still in developer preview and might change before they are stable, you can already use them within your applications. As the names of the two functions already imply, they are used to convert a Signal into an Observable or to convert an Observable into a Signal, respectively. Let's learn about both, starting with the `toSignal` function.

Using toSignal

The `toSignal` function is used to transform an Observable into a Signal. The `toSignal` function is very similar to the `ASYNC` pipe, only with more flexibility and configuration options, and it can be used anywhere in the application. The syntax is pretty straightforward; you use the `toSignal` function and supply it with an Observable:

```
counter = toSignal(this.countObs$);
```

In the preceding example, we converted `counterObservable$` into a `counter` Signal. The `toSignal` function will immediately subscribe to `counterObservable$`, receiving any values the Observable emits from that point. As with regular Signals, you can use the Signals that were created with the `toSignal` function inside computed Signals and Signal effects. When `toSignal` changes its value because the Observable emits a new value, any Signal effect or computed Signal depending on that Signal will be triggered.

The `toSignal` function will also automatically unsubscribe from the Observable, given that the `toSignal` function is used within the injection context. When you use the `toSignal` function outside the injection context or want to make it dependent on a different injection context, you can provide the `toSignal` function with an injection context, like so:

```
injector = inject(Injector);
counter = toSignal(this.countObs$, {injector: this.injector});
```

Here, we supplied the `toSignal` function with an additional configuration object in which we provided the `injector` property. There may also be scenarios where you don't want the `toSignal` function to automatically unsubscribe the Observable when the component or injection context is destroyed.

You might want to stop the Observable earlier or later, depending on your system's needs. For this scenario, you can provide the toSignal function with a manualCleanup configuration, similar to the Signal effect:

```
counter = toSignal(this.countObs$, {manualCleanup: true});
```

When you set manualCleanup to true, the toSignal function will receive values up to the point the Observable it depends on is completed. When the inner Observable has been completed, the Signal will keep returning the last emitted value; this is also the case if you don't use the manualCleanup configuration. Besides having control over the unsubscribe process of the Observable used by the toSignal function, you can also provide an initialValue configuration.

The Observable you convert into a Signal might not immediately and synchronously emit a value upon subscription. Yet Signals always require an initial value, and if one isn't provided or the value comes in asynchronously, the initial value of the Signal will be undefined. Because the initial value of the Signal is undefined, the type of the Signal will also be undefined. To prevent undefined being the initial, you can provide the toSignal function with an initial value using this syntax:

```
counter = toSignal(this.countObs$, {initialValue: 0});
```

In this scenario, the initial value will be what you have set it to be until a new asynchronous value is received through the Observable. Having undefined as your initial value might cause problems inside computed signals or Signal effects that use the Signal with the undefined value.

On the other hand, some Observables emit a synchronous value upon subscription; think of BehaviorSubject, for example. If you use an Observable that emits a synchronous value upon subscription, you also need to configure this inside the toSignal function using the requireSync configuration:

```
counter = toSignal(this.countObs$, {requireSync: true});
```

By setting the requireSync option to true, the toSignal function enforces that the initial value upon subscription is received synchronously, skipping the initial undefined value and typing the Signal as undefined. Lastly, you can configure how the toSignal function should handle errors that happen in the Observable.

By default, if the Observable throws an error, the Signal will throw the error whenever the Signal is read. You can also set the rejectErrors option to true; in that case, the toSignal function will ignore the error and keep returning the last good value the Observable emitted:

```
counter = toSignal(this.countObs$, {rejectErrors: true});
```

When you set the rejectErrors option to true, errors are handled in the same way the ASYNC pipe handles errors within Observables.

Using toObservable

The `toObservable` function is used to convert a Signal into an Observable. Here's an example of how you can use the `toObservable` function:

```
counter = toSignal(this.countObs$);
countObs$ = toObservable(this.counter);
```

When you use the `toObservable` function behind the scenes, Angular will use a Signal effect to track the values of the Signal you use inside the `toObservable` function and emit the updated values to the Observable. Because Angular uses a Signal effect to update the values of the Observable, it will only emit stabilized changes in the Signal. So, if you set the Signal multiple times in a row without an interval, the Observable that's created with `toObservable` will only emit the last value when the Signal is stabilized.

By default, the `toObservable` function uses the current injection context to create the Signal effect; if you declare the `toObservable` function outside the injection context or if you want to create the Signal effect inside another injection context, you can provide the `toObservable` function with an injection context:

```
injector = inject(Injector);
countObs$ = toObservable(this.counter, {injector: this.injector});
```

That's all there is to the `toObservable` function – there are no other configuration options; you use the `toObservable` function and provide the function with your Observable to convert the Signal into an Observable. You can also combine both the `toSignal` and `toObservable` functions in one go.

Here's an example of how you could use a Signal input and the `toObservable` and `toSignal` functions to fetch a new product and convert it into a Signal each time the component receives a new ID input:

```
id = input(0);
product = toSignal(
  toObservable(this.id).pipe(
  switchMap((id) => this.service.getProduct(id as number)),
  ),
  { initialValue: null }
);
```

Now that you know how to combine RxJS and Signals by using the `toSignal` and `toObservable` functions, let's finish this chapter by providing a bit more clarity about when to use Signals and when to use RxJS.

Choosing between Signals and RxJS

As mentioned previously, neither Signals nor RxJS are one-size-fits-all solutions. When you're building an application, chances are you'll need both Signals and RxJS to create the most optimal code. The most straightforward distinction is that Signals handle synchronous code, and RxJS is used to handle synchronous code, but things aren't always as simple. You could also convert synchronous code using the `toSignal` function. For clarity, we'll go through some examples at face value and determine if using Signals or RxJS would be better. In the real world, there are always nuances, and you should take whatever best fits your scenario, the team, and the existing code of the application.

Let's start with an HTML template. In an HTML template, you can use an RxJS Observable by using the `ASYNC` pipe, or you can use a Signal. I would try to use Signals in the HTML template as this simplifies the HTML template by maintaining a synchronous approach. Using Signals will also help improve the change detection mechanism Angular uses, which can improve your application's performance.

There are more gray areas in the component classes where it might be more complex to determine whether to use a Signal or RxJS Observable. If we look at the local component state, I would use Signals and computed Signals to define the component state; this also allows you to consume the component state as signals inside the HTML template.

When it comes to handling user events, it depends a bit on how you need to process the values of the event. If it's a simple event such as handling a form submission or a button click, a Signal will work perfectly fine to update the correlating values. If you need more control over the delivery of the value stream, combine multiple events, or map, filter, and transform the data stream before it reaches your application logic, RxJS will be a better fit.

A typical example is when you have a search input field that makes API requests. You don't want to make too many API requests by firing an API call on each key-up event. Instead, you want to check if the user stopped typing for a specified interval. Using RxJS, this can be done using the `debounceTime` operator. You can handle the same functionality using a Signal, but this requires a lot more code, and it becomes more complex and less readable. Depending on your architecture, most other scenarios that are handled inside your components are connected with your facade services or state management.

Now, let's discuss some different scenarios and compare Signals with RxJS. Events that have to be distributed throughout the application and where different parts of your application have to react differently are also best handled using RxJS, more specifically an RxJS `Subject`. Using a `Subject` class, each part of your application can listen for the Observable and react how it needs to react.

Defining simple synchronous global application states can be done using Signals and computed Signals. The current value of the state can be retrieved by using Signals. Additionally, you can define change events using RxJS subjects if different application parts need to perform different logic when the values change. You can trigger the RxJS subjects inside your state management using a signal effect. Using the Signal effect might only work if reacting on stabilized value changes is enough; if you need to react to multiple changes that follow on from each other, this approach will not work for you.

When you have more complex state or asynchronous sources that need to be modified, combined, filtered, or mapped before you can provide the values to the rest of your application, RxJS is the best solution to handle the data streams. Especially when you need to handle multiple nested Observables or if you want to combine various streams and need control over when and how the values of these different streams are processed, RxJS offers many more tools to handle this gracefully.

Inside your facade services, you can combine RxJS and Signals. Depending on the complexity and setup of your state, a good approach is to use the `toObservable` function and RxJS to create the models you need to expose to the view layer. Once you've mapped all the data streams into the models and values you need, you can use `toSignal`, Signals, and computed Signals to expose the values to the view layer. Then, inside the view layer, you can consume the Signals synchronously while the facade service updates them asynchronously.

Now that you have a better idea of when to use Signals and when to use RxJS, let's move on to the next chapter and start learning about state management.

Summary

In this chapter, you learned about reactive programming. You learned what reactive programming is, how the Angular framework uses it, and how it can be utilized to make your code efficient, event-driven, and performant.

Next, we did a deep dive into RxJS and saw how it can be used to create and handle Observable streams. You learned about different types of Observables and how to combine, flatten, and modify Observable streams using RxJS operators. We also explored some of the most used RxJS operators and learned how to create operators using the pipe function.

After understanding RxJS, we moved on to Angular Signals. You learned why Angular introduced Signals into the framework and how they help simplify your Angular code and improve the performance of your applications. You learned about Signals, computed Signals, the Signal effect, and interoperability functions for Signals and RxJS. We finished this chapter by exploring when you should use Signals and when to use RxJS within your applications.

In the next chapter, we will take a deep dive into state management.

8

Handling Application State with Grace

In this chapter, you'll learn about application state. Understanding and handling the state of your application is one of the most essential parts of frontend development. If the state of your applications becomes messy, entangled, and hard to understand, your development process and the quality of your application will suffer.

To help you better manage your application state, we will talk about the different state levels you'll find within your applications. You will learn how to partition and divide your state for maximum efficiency. You will also create a state management solution using RxJS and Signals and build a facade service to access your state from the component layer.

Next, you will learn how to handle more complex states with the NgRx library. NgRx is the most commonly used state management library within the Angular community and uses the Redux pattern to manage state. Since the introduction of Angular Signals, NgRx also provides different approaches to working with Signals while using the tools we love from NgRx.

By the end of this chapter, you will have implemented state management solutions using different methods. You'll have learned how easy it is to change your state management solution when using a facade service and seen how Angular Signals has changed the way we handle state within Angular applications.

This chapter will cover the following topics:

- Understanding application state
- Handling global application state using RxJS
- Handling global application state using Signals
- Handling global application state with NgRx

Understanding application state

In simple terms, **application state** is a snapshot of the current condition (or state) of your data, configurations, and views at a specific point in time. Application state is the sum of all actions that are performed within your application from the moment it is loaded in the browser. The state is a dynamic landscape that influences your application's view, user interactions, data flow, and overall functionality.

It's essential to have good state management within your application so that all your components can display the correct data to the end user and you have accurate data to work with within your application code. Good state management prevents unintended data changes, resulting in incorrect views and operations being performed within your application code that you did not intend to perform.

Now that you have an idea of what application state is and why you need it, let's dive deeper, starting with the different levels of application state.

Different levels of application state

In the realm of frontend development, we can distinguish between two levels of state: global and local state. In this section, we'll delve into the nuanced distinction between global and local application states within the context of Angular, shedding light on their roles in crafting robust and maintainable frontend applications.

As their names imply, **local state** is localized to a file, component, or element within your application, whereas **global state** is shared through your entire application. The global application state serves as the central repository for shared information across various components, ensuring coherence and synchronicity in the application's behavior. On the other hand, the local application state encapsulates the internal data and configuration specific to individual Angular components and services.

By understanding the dual nature of global and local states, your Angular applications can strike a harmonious balance between reusability, encapsulation, and shared data integrity. Let's start by diving a bit deeper into the local application state within Angular applications.

Local application state

When we refer to local state, we're talking about properties that have been localized to a component or service that determine how that component or service behaves and presents data to your application's user.

A simple example of a local state would be a Counter component with a count state:

```
export class Counter {
  count = signal(0);
  add() { this.count.update((count) => count + 1) }
  subtract() { this.count.update((count) => count - 1) }
}
```

The count property is used to display the current count to the user. The declaration and update behavior of the count property is handled locally within the current component.

Within components, you can consider the state as local when the stateful property isn't shared between multiple smart components and doesn't need to be persisted when you navigate from one page to another.

Within services, the state can be considered local whenever it entails a private property that isn't shared with the outside world, and the property doesn't have to persist longer than the life cycle of the service file. If the property doesn't meet these criteria, you probably need to locate it somewhere within your global application state.

Here are some common examples of local state within Angular applications:

- Disabled button state
- Form validity state
- Modal visibility
- Sorting and filtering
- Accordion state
- Selected tab state

You now have a good understanding of local state. You know what local state is, how you can recognize it, and what the preferred tool is to handle local state within your Angular applications. You also learned about some common examples of local state. Next, you will learn about global application state.

Global application state

In contrast to local state, global application state refers to the data and configurations you share across multiple components and services within an Angular application. You can think of your global application state as a centralized repository for your data. This centralized repository of information is pivotal in ensuring consistency, synchronization, and efficient communication between various application parts.

Unlike local state, which is confined to a specific component or service, global application state persists throughout the entire application, making it particularly useful for scenarios where data needs to be shared and synchronized across different components and services, as well as during the whole user session.

Within Angular application, global state is commonly handled inside services. By creating a service dedicated to managing global state, developers can ensure that components have a centralized access point to crucial information. Services that contain global application state are often named *stores*. For example, you can call the service to store global user state user.store.ts with a UserStore class.

Within smaller Angular applications, state is commonly managed using `Subjects`. More specifically, `BehaviorSubject` stores and distributes stateful properties, and the regular `Subject` distributes global events. With the introduction of Signals, some `BehaviorSubject` classes can be replaced with Signals. We will see this in detail in the *Handling global application state using RxJS* section when we start building global state management.

For larger Angular applications, libraries such as NgRx, NgXs, Akita, and Angular Query are the preferred methods for handling global state. These libraries enhance your capabilities to manage the states gracefully and implement structured and battle-tested design patterns for managing and updating the global state in a predictable and scalable manner.

Understanding when to utilize the global application state is crucial. A global state might be more appropriate if a stateful property needs to be shared across multiple smart components or persists beyond a single component's life cycle. Now that you know what local and global state are, when to use which, and what tools are available to manage both gracefully, let's learn about some important concepts within state management.

Fundamental concepts within state management

To build a robust state management system within your Angular applications, you need to understand the fundamental concepts of state management. You need to know these concepts, why they are essential, and the dangers of not using them.

In this section, we will learn about unidirectional data flow, immutability, and side effects. Other important fundamentals for state management include reactivity and design patterns such as the Redux pattern, but we already discussed both in *Chapters 6* and *7*, so we won't dive deeper into this.

Unidirectional data flow

Unidirectional data flow is the first concept of state management we will discuss. As its name suggests, the concept states that data should only flow in one direction throughout your application. Changes to the data occur through well-defined actions or events, ensuring a clear and predictable flow of information. Unidirectional data flow simplifies debugging, makes code more predictable, and enhances maintainability. It prevents unexpected side effects by enforcing a clear flow of data through the application.

Without unidirectional data flow, tracing the origin of state changes becomes challenging, leading to debugging difficulties and potential issues with data consistency. Uncontrolled data flow can result in unpredictable behavior, especially in large and complex applications.

The concept of unidirectional data flow is important throughout your entire application and for both local and global application state. For the global application state, I recommend always having a unidirectional data flow. Within the local component state, you can sometimes make an exception by using Angular two-way data binding.

To help you understand how unidirectional data flow looks in an Angular application, here's an example of flow:

1. The state is passed from the store to the facade service.

2. The state is passed from the facade service to the smart component.

3. The smart component passes the data to the (dumb) child components.

4. The view is rendered based on the state of the smart component and its child components.

5. An action can be triggered in the view.

6. The event of the action and the related data move from the (dumb) child component up to the smart component.

7. The smart component or facade dispatches an action to the store.

8. The store updates the state based on the dispatched action.

9. The state is passed from the store to the facade.

As you can see, the data starts in the store and flows in one direction until the view can be rendered. When the user triggers an action within the view, the data flows in one direction and in a predictable manner back into the store until we reach full circle. Now that you know what unidirectional data flow is and why it's important in state management, let's learn about immutability.

Immutability

Immutability involves the practice of not modifying existing data structures directly. Instead, new copies are created with the desired changes, preserving the integrity of the original data. Immutability simplifies state management by providing a single place to mutate your state. It helps prevent unintended state changes and side effects and is particularly valuable when you're tracking and managing complex state in Angular applications.

With immutability, you may find it easier to track state changes and keep your state synchronized. Directly modifying state objects can lead to bugs and unexpected behavior. Immutability was mainly used within global state management, but with the introduction of Signals, it's now also applied within the local state of Angular applications.

Side effects

Side effects refer to an operation or changes you perform when a specific piece of your state changes. Side effects can include things such as the following:

- Fetching data
- Updating the local storage
- Dispatching additional actions
- Setting local variables

By isolating side effects, you can maintain a clear separation of concerns in your application. The core application logic (reducers, actions, and selectors) remains focused on state changes, while side effects are handled separately. Side effects are natively introduced in the Angular framework within the Signals API, and they are used in popular state management libraries such as NgRx and NgXs.

So, to summarize, there is local and global state within your applications. Local state is localized to components or services, whereas global state affects the entire application. Some of the fundamental concepts of state management are unidirectional data flow, immutability, and side effects. You learned about the advantages of these concepts and why they are important for a state management solution. You also learned what state management is and why you need it within your applications.

In the next section, you will start building a global state management solution and create a facade service to access the state from within your smart components.

Handling global application state using RxJS

In this section, you will create a simple state management solution using RxJS. At the core of this state management solution lies the RxJS `BehaviorSubject` class. You will also create a facade service to interact with the state management solution.

The facade will do all the communication with the state management solution and the smart components. This decouples our smart components from the state management solution, making it easy to swap our state management implementation when needed.

Once we've created the RxJS state management solution and connected it with the component layer of the application, we can change the state management and facade to use Signals where it is possible and makes sense.

By converting the state management solution from RxJS into Signals, you'll be able to understand both concepts and learn about the differences. Building both approaches will also serve you best so that you can recognize and work with both systems when you encounter them in a project you join. Let's start by building the RxJS state management solution.

Building a state management solution using RxJS

To start building the state management solution, create a folder named `stores` in the `data-access` library of the finance domain. The `stores` folder should be located inside the `lib` folder, at the same level as the `adapters`, `HTTP`, `models`, and `services` folders.

Creating a service

First, you can create a service by using the *Nx generator* in the newly created `stores` folder. Name the new service `expenses.store`. Because we're using the Nx generator, it will create a file named `expenses.store.service.ts`; you can remove the `.service` part manually and do the same for the `spec` file.

Next, rename the class `ExpensesStore` instead of `ExpensesStoreService` and remove `constructor`; this should be in your file when you're ready:

```
@Injectable({ providedIn: 'root' })
export class ExpensesStore {}
```

Next, you need something that can hold the state for your list of expenses. We will use a `BehaviorSubject` class that will emit an array of `ExpenseModel`. The `BehaviorSubject` class will be a private property, so you cannot directly mutate the state from outside our `ExpensesStore` class.

Only the `ExpenseStore` class should be able to mutate the state directly; all other parts of the application should mutate it through `ExpenseStore` and, more precisely, through the facade, which will call `ExpenseStore`. Allowing the state to be directly mutated from other parts of your application can lead to unintended state mutations, breaking your application.

Because the `BehaviorSubject` class is private, you also need a public property that exposes the `BehaviorSubject` class to the outside world as an Observable:

```
private expenses = new BehaviorSubject<ExpenseModel[]>([]);
expenses$ = this.expenses.asObservable();
```

As you can see, we first defined the `expenses` `BehaviorSubject` class, and we created the public `expenses$` Observable by taking the `expenses` `BehaviorSubject` class and calling the `asObservable()` method on the `BehaviorSubject` class. We gave the `expenses` `BehaviorSubject` class an empty array as its default value. Next, let's add some logic to fetch and distribute our data.

Fetching and distributing data in our store

Next, we'll add some logic to make an API request that retrieves the expenses and emits the received expenses through the `BehaviorSubject` class. To achieve this, start by injecting the `ExpensesHttpService` class we created in *Chapter 6*:

```
protected expensesApi = inject(ExpensesHttpService);
```

Next, you need to create a method to make the API request and update the `BehaviorSubject` class:

```
fetchExpenses(): void {
  this.expensesApi.get().subscribe({
    next: (expenses) => { this.expenses.next(expenses) },
    error: (err) => { console.log(<err ==>>, err) }
  });
}
```

As you can see, we made a method named `fetchExpenses`, and inside this method, we used `expensesApi` to make a `get` request. We subscribed to the `get` request and handled the `next` and `error` events of the subscription. The `next` error is handled when the subscription of the `get` request receives a response, and the `error` event is handled whenever the `get` request fails and returns with an error status.

If the API request responds successfully, we call the `next()` method on the `expenses` `BehaviorSubject` class and give it the received `expenses` as a parameter. If the API responds with an error, we simply log the error. In a production application, you should handle this better and alert the user with a toaster message or something similar.

Adding additional expense methods

Next, you'll want to add methods to get an expense by ID, as well as update, delete, and add expenses. Before you create these methods, `MockInterceptor` must be adjusted to handle the `delete` and `getByID` requests. You can modify the interceptor yourself or get the adjusted `MockInterceptor` from this book's GitHub repository: `https://github.com/PacktPublishing/Effective-Angular`.

After adjusting `MockInterceptor`, you can start to implement the `add`, `delete`, `update`, and `getByID` methods in our expenses store. Inside all these methods, we need access to the current list of expenses. You can access the current list of expenses through the `value` property of the expenses `BehaviorSubject` class. Let's create a getter that retrieves the current expenses from our state:

```
private get currentExpenses() {return this.expenses.value}
```

Now, we can start adding the methods.

Adding expenses

Let's start by creating a method to add an expense:

```
addExpense(expense: ExpenseModel): void {
  this.expensesApi.post(expense).subscribe({
    next: (addedExpense) => {
      addedExpense.id = !addedExpense.id ? this.currentExpenses.length
+ 1 : addedExpense.id;
      this.expenses.next([...this.currentExpenses, addedExpense]);
    },
    error: (err) => { console.log(‹err ==>›, err) }
  })
}
```

As you can see, the addExpense code takes expense as a function parameter. This expense parameter is used to call the POST request on expenseApi.

When the API returns with the response, we update the ID property (we're only updating the ID property because we don't have an actual backend. Normally, ID would be populated by the backend). After updating the ID property, we add the newly created expense to the expenses state.

Deleting expenses

After creating the addExpense method, you can make a method to delete an expense:

```
deleteExpense(id: number): void {
  this.expensesApi.delete(id).subscribe({
    next: () => {
      this.expenses.next(this.currentExpenses.filter(expense =>
expense.id !== id));
    },
    error: (err) => { console.log(‹err ==>›, err) }
  })
}
```

The delete method is pretty straightforward. We make the API request, and when the API responds, we update the expenses state with the new list of expenses by calling the next() method. As a parameter for the next() method, we use the current list of expenses and filter out the deleted expenses. If the API responds with an error, we log the error, again in a production application, and we show some sort of message to the user.

Fetching, getting, and selecting expenses

After adding the `delete` method, we will add the `getExpense`, `selectExpense`, and `fetchExpenseById` methods. The `getExpense` and `selectExpense` methods will be public methods, while the `fetchExpenseById` will be private. We will also create an expense `Subject` class and a `selectedExpense` state using a `BehaviorSubject` class.

Let's start by adding the `Subject` class and the `selectedExpense` state:

```
private expense: Subject<ExpenseModel> = new Subject();
expense$: Observable<ExpenseModel> = this.expense.asObservable();

private selectedExpense: BehaviorSubject<ExpenseModel | null> = new
BehaviorSubject<ExpenseModel | null>(null);
selectedExpense$ = this.selectedExpense.asObservable();
```

The expense `Subject` class and `selectedExpense` state can be used to retrieve the selected expense reactively. The `selectedExpense` state is used when you need to persist the selection in your global application state. In contrast, the expense `Subject` class can be used to emit an expense as an event that is only received by Observers who are subscribed when the event is emitted. After adding the expense `Subject` class and the `selectedExpense` state, we will continue with the private `fetchExpenseById` method:

```
private fetchExpenseById(id: number, select = false) {
  this.expensesApi.getById(id).subscribe({
    next: (expense) => { select ? this.selectedExpense.next(expense) :
this.expense.next(expense) },
    error: (err) => { console.log(<err ==>>, err) }
  })
}
```

The `fetchExpenseById` method has `id` and `select` parameters. The `id` parameter is required, and the `select` property is optional with a default value of `false`. The method starts by making an API call to retrieve an expense by ID. When the API responds with an expense, we emit a new value using expense `Subject` or emit a value and set the `selectedExpense` state using the `BehaviorSubject` class. Depending on the needs of your application, you can also add the fetched expense to the `expenses` state, but for our demo application, this isn't necessary.

Now, to finish up the get expense by `id` logic, we need to implement the public `getExpense` and `selectExpense` methods:

```
getExpense(id: number): void {
  const expense = this.currentExpenses.find(expense => expense.id ===
id);
  expense ? this.expense.next(expense) : this.fetchExpenseById(id);
}
```

```
selectExpense(id: number): void {
  const expense = this.currentExpenses.find(expense => expense.id ===
id);
  expense ? this.selectedExpense.next(expense) : this.
fetchExpenseById(id, true);
}
```

As you can see, the `getExpense` and `selectExpense` methods are very similar. Both methods receive `id` as a parameter and check if the expense with the provided `id` parameter can be found inside the current `expenses` state.

When the expense is found in the current state, the `next()` method is called on the `expense` `Subject` class or the `selectedExpense` `BehaviorSubject` class. When no expense is found in the current expenses state, the `fetchExpenseById` method is called to get the expense from the backend; in that case, the `fetchExpenseById` method will call the `expense` `Subject` or `selectedExpense` `BehaviorSubject` class. Now that we've added everything to get or select an expense reactively, let's add the `updateExpense` method.

Updating expenses

The `update` method will receive `expense` as a function parameter. Next, it will make a PUT request using `expensesApi` to update the request in the backend. After the API responds successfully, the method will update the `expenses` state:

```
updateExpense(expense: ExpenseModel): void {
  this.expensesApi.put(expense).subscribe({
    next: (expense) => {
      this.expenses.next(this.currentExpenses.map(exp => exp.id ===
expense.id ? expense : exp));
    },
    error: (err) => { console.log(<err ==>>, err) }})
}
```

As you can see, we make the API request and use the `next()` method on `expenses` `BehaviorSubject` to update the `expenses` state. As an argument for the `next()` method, we take the `currentExpenses` getter and use the `map()` function to replace the updated expense.

Now that we've added the methods to add, update, delete, and get expenses, let's finish up the store with some additional state and methods to reset the state.

Extending ExpensesStore

We will start by adding an additional piece of state to manage whether we show prices, including or excluding VAT. We can do this by creating a new `BehaviorSubject` class and a method to adjust the `BehaviorSubject` class' value:

```
private inclVat = new BehaviorSubject<boolean>(false);
inclVat$ = this.inclVat.asObservable();

adjustVat(): void {
  this.inclVat.next(!this.inclVat.value);
}
```

As you can see, the VAT state is just a simple Boolean indicating whether we show the prices, including or excluding VAT.

Lastly, we need some logic to reset our application state and clear the selected product state. We will create two different methods for this `resetState` to reset all the states to the default values. We'll use a `clearExpenseSelection` method to clear the `selectedExpense` state:

```
clearExpenseSelection(): void {
  this.selectedExpense.next(null);
}
resetState(): void {
  this.expenses.next([]);
  this.selectedExpense.next(null);
  this.inclVat.next(false);
}
```

This was the last piece of the puzzle for our expense store. You created a simple yet effective state management solution to handle the global application state of the expenses. You did so using the RxJS `Subject` and `BehaviorSubject` classes. This `ExpensesStore` can now be used as the single source of truth for all your expense data throughout your application.

If a component needs the current state of some expense data, it will come from this `ExpensesStore`. When your application grows, and you have other entities with a state besides the expenses, such as users, reports, or settings, each entity will have a store file to manage the state of that entity.

Now that you've created a state management solution using RxJS, we will start building the facade service and connect the view layer with the store through the facade.

Connecting your state management and view layer with a facade service

Now that you have a state management solution in place, it's time to connect it to the view layer of your application. As mentioned several times in this book, the best approach is to create a facade service for this. This facade provides an additional layer of abstraction, providing a simple interface for your view layer to interact with the application state. *Figure 8.1* provides a visual representation of the facade service and how data flows from your state through the facade into your components:

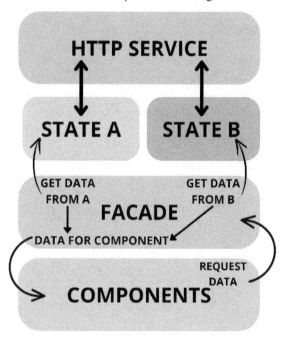

Figure 8.1: Data flow with a facade, components, and state

As you can see, your components make a simple request to the facade service, and the facade will collect the data from your different state services and send it back to the component in the format that the component needs. This ensures your components only have one dependency and your facade will host all other necessary dependencies to retrieve the data you need in your components.

Creating a facade service

Start by creating a `facades` folder inside the `lib` folder of your `expenses data-access` library. The new `facades` folder will be located in the same folder as the `store` folder.

Inside the new `facades` folder, you must create a file named `expenses.facade.ts` with an injectable class called `ExpensesFacade`. You can use the Nx generator to create a service and rename it or create the facade manually. Also, add an export inside the `index.ts` file so that you can consume the facade outside the library.

When you're finished, you should have the following in the `expenses.facade.ts` file:

```
@Injectable({ providedIn: 'root' })
export class ExpensesFacade {}
```

Creating the facade interface

Next, create a file next to the `expenses.facade.ts` file named `expensesFacade.interface.ts`. In this interface, we will declare the blueprint of our facade. So long as your facade implements this interface, you can switch the implementation of the state without touching the component layer. If you change the interface, you also need to adjust the component layer.

In the interface file, declare the following interface:

```
export interface IExpensesFacade {
  expenseSelector$: Observable<ExpenseModel>;
  selectedExpenseSelector$: Observable<ExpenseModel>;
  inclVatSelector$: Observable<boolean>;
  addExpense(expense: ExpenseModel): void;
  adjustVat(): void;
  clearExpenseSelection(): void;
  deleteExpense(id: number): void;
  fetchExpenses(): void;
  getExpense(id: number): void;
  getExpenses(id: number): Observable<ExpensesViewModel>;
  resetExpenseState(): void;
  selectExpense(id: number): void;
  updateExpense(expense: ExpenseModel): void;
}
```

After defining the interface, we can start with the facade service. Start by implementing the interface:

```
export class ExpensesFacade implements IExpensesFacade {…}
```

Now, you want to inject `ExpensesStore` inside the facade service:

```
protected readonly expensesStore = inject(ExpensesStore);
```

Now that we've injected the store, we will add a method to fetch the expenses.

Connecting the facade with the store

Let's add a simple method to the facade that simply calls the `fetch` method inside the store:

```
fetchExpenses() {
  this.expensesStore.fetchExpenses();
}
```

Next, we will create a method to get the fetched expenses. But before we do this, we will create a new interface called `ExpensesViewModel`:

```
export interface ExpensesViewModel {
  total: number;
  inclVat: boolean;
  expenses: ExpenseModel[];
}
```

You can also adjust `ExpenseModel` a bit and rename the `amountExclVat` property to `value`. If you're using VS Code, you can select the property and press *F2* to rename it. When you rename using the *F2* key, the property will be renamed in every instance (besides the HTML templates).

Now that you've created `ExpensesViewModel` and adjusted `ExpenseModel`, let's create the `getExpenses` method inside the facade:

```
getExpenses(): Observable<ExpensesViewModel> {
  return combineLatest([this.expensesStore.expenses$, this.
expensesStore.inclVat$]).pipe(
    distinctUntilChanged(),
    map(([expenses, inclVat]) => ({
      expenses: structuredClone(expenses).map(expense => {
        expense.amount.value = inclVat ? expense.amount.value * (1 +
expense.amount.vatPercentage / 100) : expense.amount.value;
        return expense;
      }),
      inclVat,
      total: expenses.reduce((acc, expense) => {
        return acc + (inclVat ? (expense.amount.value * (1 + expense.
amount.vatPercentage / 100)) : expense.amount.value);
      }, 0),
    }))
  );
}
```

As you can see, there is quite a lot going on in this method. This is one of the reasons why using a facade service is beneficial.

In a large application, the chances are high that you need this `ExpensesViewModel` in multiple components. Instead of having to define this chunk of logic in multiple component classes, you can define it inside the facade, and inside the component layer, you can use a simple function call, keeping your components simple and clean. Also, when you need to adjust the logic, you only have to do it in this single place instead of in multiple component classes. Now, to better understand what we did inside the function, let's break it down line by line:

1. We started by naming the method `getExpenses` and specified that it would return an `ExpensesViewModel` Observable.

2. Inside the `getExpenses()` method, we returned an Observable using the `combineLatest()` method.

3. Inside `combineLatest()`, we combined the `expenses$` and `inclVat$` Observables from the store and applied the RxJS `pipe()` function to `combineLatest()`.

4. Inside the `pipe()` function, we applied two operators, starting with the `distinctUntilChanged()` operators, so that we only emitted a new value when the values changed.

5. Next, we used the `map()` operator to map the two Observable streams into `ExpensesViewModel`.

6. Depending on the state of the `inclVat$` Observable, we returned the expense value properties and the total property, including or excluding VAT.

Now that you've created the `fetch-` and `getExpenses` methods inside the facade, let's adjust the expenses overview page.

Adjusting the expenses overview page

Inside the page component, start by injecting the facade service:

```
protected readonly expensesFacade = inject(ExpensesFacade);
```

After injecting the facade, you can fetch the expenses inside the `ngOnInit()` method of the page component:

```
ngOnInit() { this.expensesFacade.fetchExpenses() }
```

Next, you can clean up the component. In *Chapter 7*, we used a Signal with mocked data for expenses; in this section, we will use the expenses we receive from the `getExpenses` method inside the facade. Start by reassigning the `expenses` property, like this:

```
expenses = this.expensesFacade.getExpenses();
```

After reassigning the expenses property, you will get some errors inside the component and template file of the expenses overview page because you don't have the expenses Signal anymore. Go ahead and remove the totalInclVat computed Signal; you can also remove the Signal effect in the component and clear out the logic inside the onAddExpense method.

Next, we need to make some adjustments to the HTML template.

Start by adding an if-else block around the HTML table:

```
@if(expenses | async; as expensesVm) {......} @else {Loading... }
```

Inside the if block, you will use the expenses property with an async pipe so that you can retrieve expenses from the facade and use those values within the template.

After adding the if block, you need to adjust the for block inside the HTML template and switch the expenses Signal for the expenses property you retrieved from the facade:

```
@for (expense of expensesVm.expenses; track expense.id) {…}
```

After adjusting the for block, you need to adjust the table rows to correctly reflect the new model structure and to improve the UI. Do this by rounding the value to two decimals, and then adding a currency pipe and percentage (%) sign:

```
<td>{{ expense.amount.value.toFixed(2) | currency }}</td>
<td>{{ expense.amount.vatPercentage }}%</td>
```

Lastly, you need to switch the totalInclVat computed Signal we used inside the template for the total property on expensesVm:

```
<td>Total: {{expensesVm.total}}</td>
```

Here, we adjusted the text to total because we now show the total, including or excluding VAT, depending on the global state. After making these adjustments, you should see the total amount and expenses inside the table again, only now using RxJS and the global state instead of the Signals with mocked expenses.

Next, you want something to toggle the VAT so that you can see the expenses and total amount being automatically updated when the VAT status changes.

Start by adding a new method inside the facade service:

```
adjustVat() { this.expensesStore.adjustVat() }
```

As you can see, this is just a simple method calling the adjustVat method inside the store. This will change the inclVat BehaviorSubject class inside the store. This, in turn, will trigger the combineLatest() method we used inside the getExpenses method inside the facade.

So, when you change the VAT status, `ExpensesViewModel`, which we retrieved through the `getExpenses` method, will automatically be updated and show the total and expense amounts, including or excluding VAT, depending on the state.

Once you've added the method to adjust the VAT, you also need something to retrieve the `inclVat` status inside the facade. You can simply create a property for this and assign it with the `inclVat$` Observable from the store:

```
inclVatSelector$ = this.expensesStore.inclVat$;
```

After adding the method and property to adjust and retrieve the VAT status, let's add a toggle inside the HTML template of the expenses overview page to adjust the VAT state:

```
<div class="vatToggle">
  <span>Incl. VAT:</span>
  <label class="switch">
    <input (click)="expensesFacade.adjustVat()" type="checkbox"
      [checked]="expensesFacade.inclVatSelector$ | async">
    <span class="slider round"></span>
  </label>
</div>
```

I've added the VAT toggle next to the **add expense** and **show summary** buttons. You can get the CSS from this book's GitHub repository or add your own styling. As you can see, we used `inclVatSelector$` from the facade combined with an `async` pipe to set the `checked` property of the VAT toggle.

We've also added a `click` event to the `input` value of the toggle to call the `adjustVat` method in the facade. If you click on the toggle, you will see the expense amounts in the table and the total amount in the table summary change to include or exclude the VAT amount, depending on the VAT state.

As you might have noticed, this is a very reactive approach because everything reacts upon the state changes automatically. The code is also very performant because the updates are performed in a non-blocking manner, allowing all your code to keep running.

Now that we've implemented the `getExpenses` method and the VAT status, let's finish up the facade service.

Finishing up the facade service

For all the other methods exposed by the store, you can add simple methods inside the facade service that call them from the store, similar to what we did with the `fetchExpenses` and `adjustVat` methods.

For the `selectedExpense` and `expense` properties inside the store, you need to add a selector property inside the facade service. Because we will also map the expenses emitted by `selectedExpense` and `expense`, we will abstract the mapping behavior into a new function so that we can reuse it:

```
private mapExpense(expense: ExpenseModel, inclVat: boolean) {
   const expenseClone = structuredClone(expense) as ExpenseModel;
   expenseClone.amount.value = inclVat ? expenseClone.amount.value *
(1 + expenseClone.amount.vatPercentage / 100) : expenseClone.amount.
value;
   return expenseClone;
}
```

Next, you can adjust the mapping of the expenses inside the `getExpenses` method like this:

```
expenses: expenses.map(expense => this.mapExpense(expense, inclVat)),
```

Lastly, we will add the selector properties for `selectedExpense` and `expense`, starting with `expenseSelector$`:

```
expenseSelector$ = this.expensesStore.expense$.
pipe(withLatestFrom(this.expensesStore.inclVat$), map(([expense,
inclVat]) => this.mapExpense(expense, inclVat)));
```

As you can see, for `expenseSelector$`, we used the `withLatestFrom()` operator and didn't use `combineLatest()`. We did this because `expenseSelector$` will only emit a value as an event using the `Subject` class instead of `BehaviorSubject`. There is no state here, and we don't want the selector to emit a new value when the VAT toggle changes. We only want it to respond when the `expense` `Subject` class emits a value, and when that happens, take the current value of the `inclVat$` Observable to map the expense.

The `selector` property for `selectedExpense` will use the `combineLatest()` function to combine the `selectedExpense$` Observable and the `inclVat$` Observable, like so:

```
selectedExpenseSelector$ = combineLatest([this.expensesStore.
selectedExpense$, this.expensesStore.inclVat$]).
pipe(filter(([expense]) => !!expense), map(([expense, inclVat]) =>
this.mapExpense(expense as ExpenseModel, inclVat)));
```

For `selectedExpenseSelector$`, we used the `combineLatest()` function because the selected expense is stateful and persists in our store. We might use the selected expense in our view while we can change the VAT, and because of that, we want it to react when the VAT status changes and update the amount in the view. Because we want `selectedExpense` to be reactive on the VAT status as well, we used the `combineLatest()` operator, which triggers whenever one of the combined Observables emits a new value.

This was the last part of implementing the state management solution using RxJS. This approach to state management is commonly used within smaller Angular applications where the state isn't used in many different components and services. The solution offers good reactivity and is easy to build and understand.

Now, let's learn how to convert this state management solution so that it uses Signals instead of RxJS. Using Signals will simplify your facade service and component layer. It also allows Angular to perform better change detection. If you need to combine many data streams with tailored logic and apply modifications to them, the RxJS approach will fit your application better.

That said, using Signals is much simpler for simple state and data streams. Even if you need to combine some data streams without needing to have too much control over how this happens, the Signal approach will be best suited for you. Signals are the way to go for your state if you find yourself only using `combineLatest()` and `withLatestFrom()` and some basic operators such as `map()` and `filter()`.

Handling global application state using Signals

To convert your state management solution so that it uses Signals instead of RxJS, you must change the `BehaviorSubject` classes in the `ExpensesStore` to Signals. You still want to ensure that the state only emits a new value when it's set in the store; you don't want to be able to set the state outside of the store.

To achieve this, we will create a private `WritableSignal` and a public Signal that is read-only. You can change all the `BehaviorSubject` classes to Signals using the following syntax:

```
private expensesState = signal<ExpenseModel[]>([]);
expenses = this.expensesState as Signal<ExpenseModel[]>;
```

Here, we declared a private Signal using the `signal()` function. Declaring a Signal in this manner will create `WritableSignal`. In the line after, we created a public property and assigned it with `WritableSignal` but cast it to a `Signal` type with the `as` keyword; here, the `Signal` type is read-only. After adjusting all your `BehaviorSubject` classes, you need to change the references you had to them inside the store.

Start by removing the `currentExpenses` getter and change all instances of `this.current Expenses` to the following:

```
this.expenses()
```

Next, inside the `adjustVat()` function, change `!this.incluVat.value` to the following:

```
!this.inclVat()
```

Lastly, you need to adjust all instances where you used the `next()` method on one of your `BehaviorSubject` classes.

Here's an example of how to convert the `resetState()` function:

```
resetState(): void {
  this.expensesState.set([]);
  this.selectedExpenseState.set(null);
  this.inclVatState.set(false);
}
```

Now, change all other instances of the `next()` method to the `Subject` class and the `set()` method. That's all we need to do for `ExpensesStore`; you now have state management that uses Signals instead of RxJS `BehaviorSubject` classes. After adjusting the state, we need to change `ExpensesFacade` so that it can work with Signals instead of Observables.

Normally speaking, one of the advantages of a facade service is that it is an abstraction layer, and we don't need to touch the component layer when changing the state management solution. But in this situation, we need to adjust both the facade service and the component layer; this is because we will be changing the interface of the facade service.

In theory, we could maintain the interface and still return Observables by converting the Signals back into Observables in the service; doing so would allow you to leave the component layer untouched. However, we want to utilize the full power of these Signals and also implement them in our components so that Angular can perform better change detection and we can make our templates synchronous. To achieve this, we need to return Signals from our facade service instead of Observables, changing the interface of the facade.

We will start changing the facade by changing the interface. Replace the `getExpenses` method inside the interface with an `expenses` property and adjust the `selectedExpenseSelector$` and `inclVatSelector$` properties like this:

```
selectedExpense: Signal<ExpenseModel | null>;
inclVat: Signal<boolean>;
expenses: Signal<ExpensesViewModel>
```

After making the preceding adjustments in the interface, you can start implementing the interface inside the facade service. To implement the changes in the interface, remove the `getExpenses` method. Instead of the `getExpenses` method, you must create a computed Signal that returns the same value as the `getExpenses` method did:

```
expenses = computed<ExpensesViewModel>(() => {
  const inclVat = this.expensesStore.inclVat();
  return {
    expenses: this.expensesStore.expenses().map(expense => this.
mapExpense(expense, inclVat)),
```

```
        inclVat,
        total: this.expensesStore.expenses().reduce((acc, expense) => {
            return acc + (inclVat ? (expense.amount.value * (1 + expense.
amount.vatPercentage / 100)) : expense.amount.value);
        }, 0),
    }
});
```

As you can see, the computed Signal is very similar to the getExpenses method. The main difference is that we no longer need the combineLatest() and map() operators. We can now use the inclVat and expenses Signals inside the computed Signals.

When one of the two Signals receives a new value, the computed Signal will automatically compute a new one. The computed Signals can be seen as combineLatest() of the Signals realm. The equivalent of withLatestFrom() would be using a Signal inside the computed Signal and wrapping the Signal using the untracked() function, as we discussed in *Chapter 7*.

After adding the computed Signal, we need to implement the inclVat and selectedExpense Signals inside the facade service. This is pretty straightforward – you simply define the property and assign it with the Signal from ExpensesStore:

```
inclVat = this.expensesStore.inclVat;
selectedExpense = this.expensesStore.selectedExpense;
```

Here, we assign the property with the Signal from the store; we do not call the Signal by adding function brackets, (). We don't add these function brackets because we want to use the actual Signal inside the component layer, not the Signal value. If you were to call the Signal here and retrieve the value inside the component layer, the update behavior wouldn't work as expected, and the view wouldn't be updated when your state changes.

The last thing to do is adjust ExpensesOverviewPageComponent and its template. Inside the component class, you can adjust the expenses property and assign it with the expenses Subject class from the facade instead of the getExpenses() function:

```
expenses = this. expensesFacade.expenses;
```

Now, inside the HTML template, you need to change inclVatSelector$ to inclVat(), remove the async pipe, and change expenses with the async pipe to expenses() without the async pipe:

```
[checked]="expensesFacade.inclVat()"
@if(expenses(); as expensesVm) { ...... }
```

With the preceding changes, you've adjusted the component class and the HTML template to use Signals instead of Observables. As you can see, using the Signal approach is slightly simpler and needs fewer lines of code. It also makes your HTML template synchronous and helps Angular to perform better change detection, leading to better performance.

The flip side is that you have less control over the data streams, and it's not as easy to modify the stream before it reaches your application logic. Compared to RxJS, Signals also offers less control when you want to combine different streams of data, so depending on your needs, you can decide whether to use Signals or RxJS.

You can also make a hybrid solution and convert Observables into Signals, giving you the best of both worlds. In that case, you can use the RxJS operators you need and still consume the values as Signals in your component classes and HTML templates.

With that, you've learned how to create a state management solution using RxJS and Signals. You made a facade service as an additional abstraction layer and learned when you have to change the component layer and when you only have to change the state management layer when working with a facade service.

Both state management solutions we've created work well for small applications with relatively simple global states. The RxJS method is implemented a lot, and with the popularity of Signals, I imagine the Signal approach will be implemented a lot as well. But when you have a larger application where the state is used in a lot of components and services, you will run into issues with our current implementation. In the next section, you will learn about these issues and how to resolve them.

The problem with using RxJS or Signals for global state management

While our current state management solution is used in many applications and works well for our current application, there's a huge problem: our current global state management solution isn't immutable.

You cannot modify your `BehaviorSubject` classes or Signals from outside the store, so in that sense, it is immutable. Also, when using primitive values for your state, the state itself is immutable. However, when you're using reference objects as values for your `BehaviorSubject` classes or Signals, the state itself isn't immutable.

When you use an array or object for your state and retrieve the state through `BehaviorSubject` or `Signal`, you can modify the value of the state unintentionally. When you adjust the retrieved state object within a component or service class, the value of `BehaviorSubject` or `Signal` is also modified!

This is also the reason why we used the `structuredClone()` function inside our `mapExpenses()` function. If you remove `structuredClone()` and toggle the VAT a couple of times in the view, you will notice that the amounts keep increasing instead of adding and removing the VAT. This happens because we modify the object inside `Signal` or `BehaviorSubject` whenever we adjust it inside the facade service.

The next time we retrieve the state, it still has the adjusted values instead of the real state we expect. Relying on developers to always clone the object when it's modified is risky and not how you want it to be.

Allowing your state to be modified outside the store and without calling the `next()` or `set()` method on `BehaviorSubject` or `Signal` opens the door for unintended state changes, resulting in a corrupted state. When your state is not what you expect it to be, you can display incorrect values to the user and perform actions within your code that aren't intended.

For small applications where the state isn't used in many places, this might be a manageable problem, but when your application grows, your state is used in multiple places and often the retrieved state is modified locally, so the problem will surface quickly.

To have a state management system that is truly immutable, reactive, and can handle the state of any application no matter how large it becomes, your best bet is to go with a good library that focuses on state management. Some popular choices within the Angular community are as follows:

- NgRx
- NgXs
- RxAngular
- Akita
- Angular Query

All of these libraries have their advantages and disadvantages. My personal favorites are RxAngular, NgXs, and NgRx. NgRx is by far the most commonly used state management solution within the community and offers support for Observable and Signal-based state management. RxAngular is gaining a lot of popularity and has a very intuitive approach for reactively managing state with little to no boilerplate code; it also enables you to ditch ZoneJS and boost the performance of your application.

In the next section, we will convert our state management solution into an NgRx state management solution. I've picked NgRx because this is the most commonly used solution, but I recommend that you investigate some of the other solutions.

Handling global application state with NgRx

When working on enterprise software or an application with extensive or complex state management, you should use a battle-tested state management solution that provides true immutability, unidirectional data flow, and good tools to perform side effects and modify the state securely. The best way forward is to use a battle-tested library that focuses on state management.

The most commonly used library for state management within the Angular community is **NgRx**; it has a huge community and all the tools you might need to handle even the most complex state. NgRx implements the Redux pattern and consists of four main building blocks: actions, reducers, selectors, and effects.

In this section, we will change our custom-made state management solution so that it uses NgRx. We will keep the store file we made in the previous section for reference purposes and build the NgRx state management in new files.

In a production environment, you should remove the old unused store file. Inside the facade service, we will simply replace the current implementation with the NgRx implementation, and this time, we will not adjust the `IExpensesFacade` interface, meaning we do not have to change our component layer. Let's go over the step-by-step process of implementing NgRx state management.

Installing the @ngrx/store and @ngrx/effects packages

To start implementing NgRx state management, you need to install some packages by running the following npm commands in the root of your *Nx monorepo*:

```
npm install @ngrx/store --save
npm i @ngrx/effects
```

After installing the `@ngrx/store` and `@ngrx/effects` packages, you need to create some folders and files. There is an Nx generator to create the initial setup for your NgRx store, but we will set up everything manually so that you get a better understanding of how everything works and what's needed when using NgRx.

Start by creating a folder named `state` inside the `lib` folder of the expenses data-access library (next to the `stores` folder). Inside the newly created `state` folder, create another folder named `expenses`. Now, inside the newly created `expenses` folder, create these five files:

- `expenses.actions.ts`
- `expenses.reducers.ts`
- `expenses.selectors.ts`
- `expenses.effects.ts`
- `index.ts`

When you're done creating the folders and files, you can start adding some actions inside the `expenses.actions.ts` file.

Defining your first NgRx actions

Actions are one of the main building blocks of NgRx and the Redux pattern. Actions define the unique events you can dispatch and perform for your NgRx state. Actions are defined as constants using the `createAction()` function, which the `@ngrx/store` package exposes to you.

You must provide the `createAction()` function with a description of the action and, optionally, with a `props()` function to define the properties you have to provide the action with to perform the action.

Alternatively, you can use the `createActionGroup()` function to create multiple events and group them under a single constant. We won't use the `createActionGroup()` function, but you can always read about it yourself in the official NgRx documentation: `https://ngrx.io/docs`.

We'll start with a simple task: defining an action to fetch the expenses from the API. You don't have to provide any argument to fetch the expenses, so the action will only contain a description. The descriptions of NgRx actions commonly use the following naming conventions:

```
[Unique State Name] Description of the action
```

Inside the `expenses.actions.ts` file, define the action to fetch expenses, like so:

```
export const fetchExpenses = createAction(`[Expenses] Fetch
Expenses`);
```

Commonly, when defining NgRx actions that include API requests, you also define a success and failure action. So, go ahead and define an action for when fetching the expenses succeeds or fails:

```
export const fetchExpensesSuccess = createAction(`[Expenses] Fetch
Expenses Success`, props<{ expenses: ExpenseModel[] }>());
export const fetchExpensesFailed = createAction(`[Expenses] Fetch
Expenses Failed`);
```

Here, we declared two actions; both received a description, and the `fetchExpensesSuccess` action also received the `props()` function. Inside the arrow brackets, we defined the type of the `props()` function – in this case, an object with an `expenses` property containing an `ExpenseModel` array. The `fetchExpensesSuccess` action needs expenses as `props()` because we will use the `fetchExpensesSuccess` action to update the state with the expenses that are retrieved from the API request.

Now that you've added the `fetchExpenses`, `fetchExpensesSuccess`, and `fetch ExpensesFailed` actions, let's update the `index.ts` file inside the `state/expenses` folder by defining an export for our expense actions:

```
export * as ExpenseActions from './expenses.actions';
```

After adding the export to the `index.ts` file, we can move on to the next piece of the puzzle. The next step is to create an NgRx effect that will make an API request to fetch the expenses and dispatch the success of the failed action accordingly.

Creating your first NgRx effect

You will create your **NgRx effects** inside the `expenses.effects.ts` file. Effects allow you to perform side effects when an action is dispatched. Effects are commonly used for tasks such as fetching data, dispatching other events, or updating local storage. Side effects isolate some logic away from your components, allowing you to keep the component classes as simple as possible.

The first effect you will create is the `fetchExpeses$` effect. This effect will run whenever the `fetchExpenses` action is dispatched. The effect will then make an API request to fetch the expenses and map the result of the API call into a newly dispatched action – the `fetchExpensesSuccess` or `fetchExpensesFailed` action.

To get started, create an injectable class inside the `expenses.effects.ts` file named ExpensesEffects:

```
@Injectable({ providedIn: 'root' })
export class ExpensesEffects {}
```

After creating the `ExpensesEffects` class, you need to inject the `Actions` class from `@ngrx/ effects` and `ExpensesHttpService` inside your `ExpensesEffects` class:

```
private readonly actions = inject(Actions);
private readonly expensesApi = inject(ExpensesHttpService);
```

Next, use the `createEffect()` function that's exposed by `@ngrx/effects` to create your first effect:

```
fetchExpeses$ = createEffect(() =>
  this.actions.pipe(
    ofType(ExpenseActions.fetchExpenses.type),
    switchMap(() => this.expensesApi.get().pipe(
      map((expenses: ExpenseModel[]) => ExpenseActions.
fetchExpensesSuccess({ expenses })),
      catchError(() => of(ExpenseActions.fetchExpensesFailed())))
```

```
        ))
    )
  );
```

In the preceding code snippet, you created your first effect named `fetchExpenses$`. As you can see, there is quite a lot going on there, so let's break it down line by line.

We started by defining a property named `fetchExpenses$` and assigned it with the `createEffect()` function. Inside the `createEffect()` function, we defined a `callback` function that returns the `this.actions.pipe()` method. The `this.actions` instance refers to the `Actions` class we injected in the previous code block. The `Actions` class emits the actions we dispatched and extends the Observable class, meaning you can use the RxJS `pipe()` function on the class.

Inside the `pipe()` function's chained-on actions, we defined a couple of operators, starting with the `ofType()` operator. The `ofType()` operator is a filter operator that filters out actions by the action type. Inside the function brackets of the `ofType()` operator, you defined the type of an action. In our case, we provided it with the type of our `fetchExpenses` action. Here, `ExpenseAction` is used to export and import our actions, `fetchExpenses` is the property name we gave our action, and `type` is a property that's exposed on all the actions we created with the `createAction()` function.

Whenever the `fetchExpenses` action is dispatched, we will move on to the next operator inside the `pipe()` function of our effect. The next operator is the `switchMap()` operator, which is used to flatten the additional Observable stream that was created by the HTTP request to fetch the expenses.

Inside the callback of the `switchMap()` operator, we made the HTTP request and added an additional `pipe()` function to the HTTP request. Inside the `pipe()` function of the HTTP request, we used the `map()` operator to map a successful HTTP response to the `fetchExpensesSuccess` action, and we provided the `fetchExpensesSuccess` action with the expenses that were retrieved from the API response. If the API request fails, we use the `catchError` operator to map it to the `fetchExpensesFailed` action.

The `createEffect()` function will automatically dispatch the returned action; this is why we don't have to call the `dispatch()` function explicitly but simply return an Observable with the action we want to dispatch. In our case, this is the `fetchExpensesSuccess` or `fetchExpensesFailed` action.

Lastly, you need to export the effects inside the `index.ts` file inside the `state/expenses` folder:

```
export * from './expenses.effects';
```

Now that we've defined the actions and created an effect to handle the `fetchExpenses` action and dispatch the `fetchExpensesSuccess` and `fetchExpensesFailed` actions, let's cover the next building block of our NgRx state by creating our state and reducer functions.

Creating your initial state and first reducer functions

Now that you've created some actions and your first effect, you need a state to perform these actions on and **reducers** to adjust the state. In NgRx and the Redux pattern, reducers are responsible for adjusting your state properties. Inside your `expenses.reducer.ts` file, you will define your initial state object and reducers to adjust the state when actions are dispatched.

Start by creating a new interface for your state object inside the `expenses.interface.ts` file:

```
export interface ExpensesState {
  expenses: ExpenseModel[];
  selectedExpense: ExpenseModel | null;
  isLoading: boolean;
  inclVat: boolean;
  error: string | null;
}
```

After creating the interface, you can create your initial state object inside the `expenses.reducer.ts` file:

```
export const initialExpensesState: Readonly<ExpensesState> = {
  expenses: [],
  selectedExpense: null,
  isLoading: false,
  inclVat: false,
  error: null
};
```

After defining the interface and initial state object, you can create the reducer by using the `createReducer()` function. The `createReducer()` function takes in your initial state as a parameter and reduces your state based on dispatched actions.

Let's start by defining the reducer function and providing it with the initial state:

```
export const expensesReducer =
createReducer<ExpensesState>(initialExpensesState);
```

In the preceding code snippet, we created a property named `expensesReducer` and assigned it to the `createReducer()` function. Inside the arrow brackets, we provided the type the reducer will modify; in our case, this is the `ExpensesState` interface. Inside the function brackets, we provided the initial state object, `initialExpensesState`.

Next, you need to add functions inside the `createReducer()` function to update the state when an action is dispatched, starting with the `fetchExpenses` action. To update the state, you must define an `on()` function and provide the `on()` function with a reference to the action it needs to react on, as well as a `callback` function to modify the state:

```
createReducer<ExpensesState>(
  initialExpensesState,
  on(ExpenseActions.fetchExpenses, (state) => ({
    ...state,
    isLoading: true
  }))
)
```

Here, we've added an `on()` function underneath the initial state object inside the `createReducer()` function. We provided the `on()` function with `ExpenseActions.fetchExpenses` so that it reacts when `fetchExpenses` actions are dispatched.

After the reference to the action, we declare a `callback` function to modify the state. Inside the function brackets of the `callback` function, you can define a parameter that will be populated with the current state object for you; it's the convention to name this parameter `state`.

Lastly, we return a new state object by spreading the current state into the object and setting the state properties we want to change. In the case of the `fetchExpenses` action, we only want to set the `isLoading` state property to `true`.

Next, we can add the reducer function for the `fetchExpensesSuccess` and `fetchExpensesFailed` actions underneath the `reducer` function for the `fetchExpenses` action:

```
on(ExpenseActions.fetchExpensesSuccess, (state, { expenses }) => ({
  ...state,
  isLoading: false,
  expenses,
  error: null
})),
on(ExpenseActions.fetchExpensesFailed, (state) => ({
  ...state,
  isLoading: false,
  error: <Failed to fetch expenses!>
})),
```

Here, we've declared two more `on()` functions and provided them with the `fetchExpensesSuccess` and `fetchExpensesFailed` actions. Inside the function brackets of the `callback` function of the `fetchExpensesSuccess` action reducer, we used destructuring to extract the `expenses` object from the dispatched action. As you might remember, you defined the `fetchExpensesSuccess` action to take the expenses that were fetched from the API request as a parameter.

Next, inside the `callback` function, we updated the `expenses` property of the state, set `isLoading` to `false`, and set `error` to `null`. If we fetch the `expenses` property successfully, there will be no errors to show to the user.

For `fetchExpensesFailed`, we don't provide parameters when we dispatch the action, so we only provide the state object in the callback, just like we did with the `fetchExpenses` action reducer. Inside the callback of the `fetchExpensesFailed` reducer, we set `isLoading` to `false` and set an error message.

With that, you've created your initial state and a `reducer` function for each of the actions you defined. When the `fetchExpenses` action is dispatched, you use a `reducer` function to set the `isLoading` state to `true`. When you're done fetching, and the `fetchExpensesSuccess` or `fetchExpensesFailed` action is dispatched, you use a `reducer` function to set the `isLoading` state to `false` and update the `expenses` or `error` state accordingly. You can use the `isLoading` state to show a spinner, the `error` state to show an error message, and `expenses` to display your list of expenses.

Now, underneath `expensesReducer`, you need to define a unique key for the expenses state:

```
export const expensesFeatureKey = 'expenses';
```

As a last step, you need to add the reducer file inside the `index.ts` file:

```
export * from './expenses.reducer';
```

After exporting the file inside the `index.ts` file, your reducer file is ready. Before moving on to the last building block of NgRx state management, which is selectors, we will add our reducer to the `ApplicationConfig` object of our `expenses-registration` app. Inside the `app.config.ts` file, add the following inside the `providers` array:

```
provideStore(),
provideState({ name: expensesFeatureKey, reducer: expensesReducer }),
```

In the preceding code, we added the `provideStore()` function and the `provideState()` function inside the `providers` array. Inside the `provideState()` function, we added an object with a name and a `reducer` property. The name receives the unique key we provided inside the reducer file and the `reducer` property receives the `expensesReducer` function.

Now that you've created the reducer and added the configuration inside the `ApplicationConfig` object, it's time to move on to the last part of our NgRx state: selectors.

Defining NgRx selectors

Selectors are the last building block of our NgRx state solution. They are used to retrieve the parts of the state you're interested in. We'll start by defining a selector that retrieves the entire expenses state:

```
export const selectExpensesState =
createFeatureSelector<ExpensesState>(expensesFeatureKey);
```

Here, we used a `createFeatureSelector()` function and provided it with the key we declared inside the `expenses.reducer.ts` file. Next, we can define additional selectors with the `createSelector()` function to retrieve specific parts of the expenses state:

```
export const selectExpenses = createSelector(selectExpensesState,
(state) => state.expenses);
export const selectError = createSelector(selectExpensesState, (state)
=> state.error);
export const selectIsLoading = createSelector(selectExpensesState,
(state) => state.isLoading);
```

In the preceding code snippet, we declared three additional selectors – one to retrieve the `expenses` state, one to retrieve the `error` state, and one to retrieve the `isLoading` state. To finish the selectors, let's export the file inside the `index.ts` file:

```
export * as ExpenseSelectors from './expenses.selectors';
```

After adding this `export`, also export the `index.ts` file from your `state` folder; this can be found in the `index.ts` file of the `data-access` library:

```
export * from './lib/state/expenses/index';
```

Now that we have all the parts of our NgRx state management system in place, it's time to adjust the facade services.

Adjusting the facade service so that they use NgRx state management

We will adjust the `fetchExpenses` method and the `expenses` Signal inside the facade service. We haven't created actions, effects, reducers, and selectors for all the other properties yet. To convert the facade service, we need to start by injecting the `Store` class, which is exposed to you by the `@ngrx/store` package:

```
protected readonly store = inject(Store);
```

After injecting the `Store` class, we can adjust the `fetchExpenses` function in the facade service. Simply remove `this.expensesStore.fetchExpenses()` inside the `fetchExpenses` function and dispatch the `fetchExpenses` action:

```
this.store.dispatch(ExpenseActions.fetchExpenses());
```

Here, you used the `Store` class and called the `dispatch()` function on it to dispatch an action. After adjusting the `fetchExpenses()` method, it's time to adjust the `expenses` computed Signal.

Inside this computed Signal, we're using the `expenses` `Subject` class from the store. We need to change this for a Signal based on `expenses` from your NgRx state.

To adjust the `expenses` computed Signal, you need to create a new property to retrieve the `expenses` state from the NgRx state and transform it into a Signal.

We can retrieve `expenses` from the NgRx state through the `selectExpenses` selector using the `select()` method on the `Store` class. Using the `select()` method on the `Store` class combined with our selector will return the `expenses` state as an Observable, so we need to use the `toSignal()` function to transform it into a Signal:

```
expensesSignal = toSignal(this.store.select(ExpenseSelectors.
selectExpenses), { initialValue: [] });
```

Now that we have the `expenses` state from the NgRx state as a Signal inside the facade service, we can adjust the `expenses` computed Signal so that it uses `expenses` from the NgRx state instead of `expenses` from the store. Simply replace the `this.expensesStore.expenses()` instances inside the computed Signal with `this.expensesSignal()` and that's it.

With that, you've changed everything you need to change and are fetching and retrieving the `expenses` state through the NgRx actions and state. Before moving on, let's add one more piece of NgRx state together so that you can understand everything that's going on in NgRx state management.

Adding additional actions, effects, reducers, and selectors

To get a better grasp of the NgRx state management we've built, let's extend it a bit with additional actions, effects, reducers, and selectors.

We will start by adding an action to adjust the `inclVat` state, just like we did before, by adding an action. Because the `inclVat` state only entails a state change and no HTTP request, you only need an action to adjust the `inclVat` state and no success and failed actions because you aren't making an HTTP request that can succeed or fail. The action to adjust the `inclVat` state also doesn't need a parameter because we will simply change the state to what it currently isn't.

You can simply create an action and provide it with a description:

```
export const adjustVat = createAction(`[Expenses] Adjust incl vat`);
```

No effect is needed for the `inclVat` state change because you don't perform an HTTP request or need to dispatch additional actions. However, you do need a new reducer function inside the `expensesReducer` to adjust the state object.

Inside the `createReducer()` function of `expensesReducer`, add an additional `on()` function to change the `inclVat` state when the `adjustVat` action is dispatched:

```
on(ExpenseActions.adjustVat, (state) => ({
  ...state,
  inclVat: !state.inclVat
})),
```

As you can see, upon dispatching the `adjustVat` action, we will change the `inclVat` state to what it currently isn't. After adding the `reducer` function, you need to add a selector to retrieve the `inclVat` property from the state object:

```
export const selectInclVat = createSelector(selectExpensesState,
(state) => state.inclVat);
```

Now, the only thing that is left to do is adjust the facade service and use the `inclVat` property from the NgRx state instead of the Signal in `expenses.store.ts`.

To adjust the facade service, start by adding an `inclVat` property and use the `toSignal()` function to convert the `selectInclVat` selector into a Signal:

```
inclVat = toSignal(this.store.select(ExpenseSelectors.selectInclVat),
{ initialValue: false });
```

After adding the `inclVat` property, all you have to do is change `this.expensesStore.inclVat()` to `this.inclVat()` inside the expenses computed Signal.

Lastly, you need to adjust the `adjustVat()` function inside the facade service. Remove what is currently in the function and replace it with a dispatch of the `adjustVat` action:

```
this.store.dispatch(ExpenseActions.adjustVat());
```

After adding the preceding code, you've made all necessary changes and you're now using the `inclVat` property from the NgRx state instead of the Signal from `expenses.store.ts`. Now, all you need to do is add the rest of the actions, effects, reducers, and selectors so that you can remove the store from the facade service entirely and use the NgRx state for everything.

As an exercise, you can try to add the additional actions, effects, reducers, and selectors yourself based on what we did for the list of expenses. After adding the additional actions, effects, reducers, and selectors, you should be able to fully adjust the expenses facade and remove the store implementation completely. If you get stuck or simply want to copy the code, you can get it from this book's GitHub repository: `https://github.com/PacktPublishing/Effective-Angular`.

In this section, you explored NgRx and learned how to use it to manage the state of your application. We discussed the default NgRx implementation to manage the state. Note that the library has more solutions and packages to offer, but this is outside the scope of this book.

Some of the other things NgRx has to offer are `signalStore` and `signalState`, two solutions that you can use to manage your state using NgRx and signals without having to convert Observables using `toSignal()`, which is what we did in this section. There are useful RxJS operators in the NgRx library. We only used the `ofType()` operators, but NgRx offers more utility operators, such as `concatLatestFrom()` and `tapResponse()`.

NgRx also offers solutions to manage component state and to dispatch actions of access state on route changes. I highly recommend exploring NgRx and other state management libraries on your own.

Summary

You've learned a lot in this chapter and brought everything we learned in *Chapter 7* together. You learned what state management is and why you need a good state management solution. You also learned about immutability, unidirectional data flow, and side effects. After some theory, you started building a state management solution using RxJS's `BehaviorSubject` and `Subject` classes.

When you finished building the state management solution using RxJS, you created a facade service that connects your component layer to the data-access and state management layers of your application. To end your custom state management solution, you converted the RxJS state implementation into a Signals implementation, further simplifying your component layer and facade service.

Finally, you learned about the shortcomings of using RxJS and Signals for your state management solution and replaced them with an NgRx implementation that uses actions, effects, reducers, and selectors.

In the next chapter, you'll learn how to improve the performance and security of your Angular applications.

Part 3: Getting Ready for Production with Automated Tests, Performance, Security, and Accessibility

In the last part, you'll learn how to improve the performance of your Angular applications and make them more secure and accessible for everyone. Starting with performance, you'll do a deep dive into Angular's change detection mechanism, learning how Angular detects changes and what actions you can take to reduce the number of change detection cycles. When you know how change detection works in detail, you'll learn how to prevent other factors from impacting the performance of your Angular applications. Then, you will explore some common security risks when developing Angular applications and how to mitigate them. Furthermore, you'll dive into accessibility, making your application content translatable using Transloco and learning how to develop applications accessible to users from different locations and abilities. Additionally, you'll learn how to write and run unit tests using Jest, and end-to-end tests using Cypress, giving you the confidence to deploy your changes without breaking anything. Finally, you'll make some final improvements, learn how to analyze and optimize your bundle sizes, and automate your deployment process.

This part includes the following chapters:

- *Chapter 9, Enhancing the Performance and Security of Angular Applications*
- *Chapter 10, Internationalization, Localization, and Accessibility of Angular Applications*
- *Chapter 11, Testing Angular Applications*
- *Chapter 12, Deploying Angular Applications*

Enhancing the Performance and Security of Angular Applications

In this chapter, you'll learn how to improve the performance and security of your Angular applications. You will do a deep dive into the change detection mechanism of Angular so you will know how you can reduce the number of components Angular has to check for changes and re-render in the browser. Next, you will learn about actions you can take to optimize the page load times and runtime performance of your Angular applications. Once you know how to enhance your Angular applications' performance, you will learn about security. You will learn what risks you can encounter when building Angular applications and how you can mitigate these risks so you can build safe applications for your end users.

This chapter will cover the following topics:

- Understanding Angular change detection
- Enhancing the performance of Angular applications
- Building secure Angular applications

Understanding Angular change detection

For small applications, performance isn't usually a bottleneck. Still, when applications grow and you start to add and compose more components, your application can become slow, impairing the user experience and decreasing user retention. One of the reasons your applications become slow is that Angular will check more and more components for changes if you develop without taking measures to help Angular perform better change detection. So, to build performant Angular applications, you need to understand how the **change detection mechanism** works so you can reduce the number of components the framework has to check for changes and re-render in the browser.

To better understand the problem, you must first understand how Angular performs change detection and where the problems start.

Let's say you have a simple component with a title property and a changeTitle() function, like so:

```
title = 'Some title';
changeTitle(newTitle) { this.title = newTitle }
```

If you call the changeTitle() function, Angular can keep everything synchronized after changing the title. Inside the call stack, Angular will first call the changeTitle() function, and all subsequent functions will be called due to the changeTitle() function being called. Then, behind the scenes, Angular will call a tick() function to run the change detection. The change detection will run for the entire component tree because you might change a value inside a service that is used in one or more components inside the component tree. This scenario will work as expected; although Angular must check the entire component tree, it will keep the application state and the view synchronized.

Now, imagine you run some asynchronous code before updating the title property; the problems will start in this scenario. Angular will detect that changeTitle() has been called and run the change detection. Because of how the call stack works, Angular will not wait for the asynchronous operations to complete before calling a function to run the change detection behind the scenes. As a result, Angular will run change detection before you update the title property, resulting in a broken application because the old value is still shown.

> **Important note**
> In reality, asynchronous changes don't break the synchronization between the code and view because Angular uses Zone.js to tackle this issue!

Now you know that asynchronous changes can lead to undetected changes. Next, let's learn about Zone.js and how Angular uses it to account for this issue so it can perform successful change detection for synchronous and asynchronous changes.

Zone.js and Angular

Zone.js has been used by Angular since Angular 2.0. The Zone.JS library monkey patches (i.e., dynamically updates the behavior at runtime) the browser API and allows you to hook into the life cycle of browser events. This means you can run code before and after browser events happen. Using Zone.js, you can create a **Zone** and run code before the code inside the Zone is executed and after all code within the Zone has finished, including asynchronous events.

To demonstrate this, here is a simple example of such a Zone:

```
const zone = Zone.current.fork({
  onInvokeTask: (delegate, current, target, task, applyThis,
applyArgs) => {
    console.log(<Before zone.run code is executed');
    delegate.invokeTask(target, task, applyThis, applyArgs);
    console.log('After zone.run code is executed');
  }
});

zone.run(() => {
  setTimeout(() => {
    console.log(<Hello from inside the zone!>);
  }, 1000);
});
```

In the preceding code, a Zone is created, and inside the Zone, we perform asynchronous code – in our case, a `setTimeout` function. The preceding code will first log the message we declared before the `delegate.InvokeTask()` method. Next, it will run the code we declared inside the `zone.run()` callback function; this can be both synchronous and asynchronous code. Lastly, when the code inside the callback is finished, the message we declared after the `delegate.InvokeTask()` method will be logged.

Behind the scenes, Angular uses a similar approach to our Zone example to create a Zone wrapped around our entire application called the **NgZone**. The NgZone includes an Observable named `onMicrotaskEmpty` that emits a value when no more microtasks are in the queue. Angular uses this `onMicrotaskEmpty` Observable to determine when all synchronous and asynchronous code within the NgZone is finished, and Angular can safely run change detection without potentially missing changed values.

In *Figure 9.1*, you can see an illustration of how the NgZone created by Angular wraps around the entire component tree, allowing Angular to safely monitor asynchronous changes:

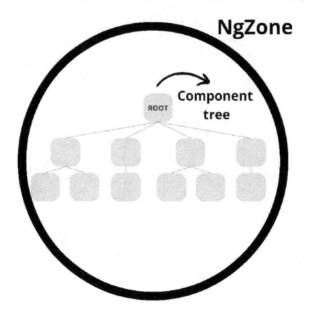

Figure 9.1: Component tree inside NgZone

When running change detection, Angular will check all components in the component tree and update and re-render the components if any bindings are changed (bindings are values bound to the HTML template).

Now you know how Angular uses Zone.js to trigger change detection when all synchronous and asynchronous tasks are finished, and that Angular checks the entire component tree when change detection runs. Let's learn why Angular checks the entire component tree and how you can reduce the number of components Angular has to check and re-render when change detection runs.

Improving change detection efficiency

Angular marks components as **dirty** when bindings or values change inside the components. When a component is marked as dirty, it is indicating the component must be updated and re-rendered. Even though components can be marked as dirty, by default, Angular will check both dirty and non-dirty components when it runs change detection. You can improve this by using the OnPush change detection strategy in your components, like in the following code:

```
@Component({
  changeDetection: ChangeDetectionStrategy.OnPush
})
```

When using the OnPush change detection strategy, Angular only runs change detection for components marked as dirty, significantly reducing the number of components that must be checked and re-rendered. There are a couple of things that mark a component as dirty:

- Browser events handled inside the component (hover, click, keydown, etc.)
- Changed component input values
- Component output emissions

When a component is marked as dirty, Angular will also mark all ancestors of the component as dirty. In *Figure 9.2*, you can see this visualized to better grasp the concept:

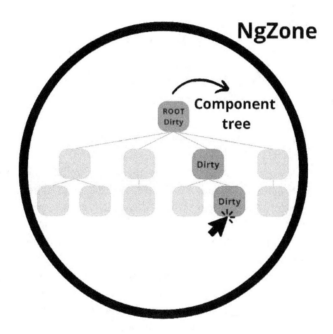

Figure 9.2: Dirty component tree

Now, components that use the OnPush change detection strategy and aren't dirty will not be checked for changes when Angular runs change detection, reducing the number of components that must be checked by the framework. Angular will also skip all child components of the components that use OnPush and aren't marked as dirty.

Figure 9.3 illustrates the mechanism of change detection with the OnPush strategy:

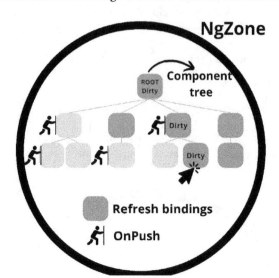

Figure 9.3: Change detection with the OnPush strategy

As you can see in *Figure 9.3*, Angular will not check whether it must refresh the bindings for all child components of non-dirty components using the OnPush change detection strategy. This also illustrates why all ancestor components must be marked as dirty when using OnPush. Angular checks whether it must refresh bindings from the top down, starting at the root component. So, if you click on a component at the bottom of the component tree, Angular starts at the root and works its way down the component tree. If the parent of the clicked component uses the OnPush change detection strategy, Angular will skip the children if the component isn't marked as dirty. As a result, Angular wouldn't check the component where you click, resulting in a mismatch between the code and view because the changes related to the click will not be processed. Because of the aforementioned reason, Angular must mark all parent components as dirty when the components use OnPush change detection.

Another interesting case for OnPush change detection is Observables. Observables are the primary tool within the Angular framework to handle asynchronous events and data streams, yet Observables receiving new values will not mark the component as dirty. So, when using the OnPush change detection strategy, the component will not be updated if the Observable receives a new value. To tackle this issue, you can use the async pipe because the async pipe will automatically mark the component to be checked and handle the update like a regular event, marking the component as dirty and running change detection afterward. Alternatively, you can use ChangeDetectorRef and call the markForCheck() or detectChanges() method manually, like so:

```
cd = inject(ChangeDetectorRef)
this.cd.markForCheck();
this.cd.detectChanges();
```

The `markForCheck()` method will mark the component to be checked during the next change detection cycle, whereas `detectChanges()` will mark the component as dirty and immediately trigger change detection for that specific component.

However, you have to be careful when using the `detectChanges()` method because it can also cause performance issues. The `detectChanges()` method will run the entire change detection in a single browser task, blocking the main threat until that task is completed. When, for example, you display a large array on the screen and must frequently detect change for this array, that results in a lot of work for the browser, slowing down your Angular application.

Now that you better understand how `OnPush` change detection works and how you can mark components as dirty or run change detection manually, let's learn how Signals are handled by the Angular change detection mechanism.

Angular change detection and Signals

In Angular 17, **Signals** was released as a developer preview, and with that, the change detection mechanism received an upgrade. When using a signal inside the template, Angular will register an effect that listens to the signal used in the template. When the value of the signal changes, the effect runs and will mark the component to be checked by Angular change detection.

The change detection cycle will work differently when the component is marked to be checked because a signal value changed. First, the component with the signal change will receive a `RefreshView` flag. Next, it will traverse up the component tree and mark all its ancestors with `HAS_CHILD_VIEWS_TO_REFRESH`. It will not mark ancestors as dirty. Now, when change detection runs, Angular will perform so-called **glo-cal (global + local)** change detection.

When running glo-cal change detection, the component tree will be checked top-down, just as usual. But when Angular encounters a non-dirty `OnPush` component with the `HAS_CHILD_VIEWS_TO_REFRESH` flag, it will skip the `OnPush` component but continue down the component tree to look for the component with the `RefreshView` flag. As a result, only the component with the `RefreshView` flag will be updated and re-rendered; all its parent components with the `OnPush` change detection strategy will not be checked or re-rendered, further improving the efficiency of the Angular change detection mechanism.

You now know how Angular change detection works and how to use the `OnPush` change detection strategy to make the change detection process more efficient. Then, you learned how to handle Observables when using `OnPush` change detection for your components. You also know how to manually mark components to be checked or run change detection using the `markForCheck()` and `detectChanges()` functions. Lastly, you've seen how you can improve change detection even more by using Signals combined with the `OnPush` change detection strategy and trigger glo-cal change detection. All these changes will significantly enhance the performance of your application, especially when your application grows and you have large and complex component trees.

In the next section, we will explore other methods to enhance the performance of your Angular applications.

Enhancing the performance of Angular applications

Understanding how Angular change detection works using the OnPush change detection strategy on as many components as possible and using Signals to further improve the change detection is a good first step to building a performant application. However, the framework has more to offer when developing performant applications.

In this section, we will explore the built-in tools and learn tips you can use to enhance the performance of your Angular applications and ensure swift page loads and good runtime performance. The first built-in tool to enhance performance that we will explore is the runOutsideAngular() method.

Understanding and using the runOutsideAngular() method

In Angular applications, optimizing performance sometimes involves executing specific tasks outside the Angular Zone. In the previous section, you learned about Zone.js, how Angular uses it to create the NgZone, and how it relates to change detection and the update behavior of your application. The runOutsideAngular() method provides a way to run specific code outside Angular's change detection mechanism, which can improve the responsiveness and efficiency of your application.

By executing tasks outsideAngular's Zone using runOutsideAngular(), you can prevent unnecessary change detection cycles from being triggered. This can lead to smoother user interactions and reduce the overhead associated with Angular's change detection mechanism. Tasks executed outside the Angular Zone are not automatically detected by Angular's change detection cycle, improving the overall performance of your application.

The runOutsideAngular() method is provided by Angular's NgZone service. The runOutsideAngular() method can run heavy computational functions outside the NgZone. Some examples of heavy computational functions are complex mathematical computations, sorting large arrays, and processing large datasets. Other scenarios in which you might want to run something outside the NgZone are as follows:

- **Running code from third-party libraries**: Running code related to initializing, configuring, or interacting with third-party libraries outside the Angular zone prevents Angular from performing unnecessary change detection, leading to better performance and avoiding potential side effects.

- **Handling WebSocket communication or long-polling requests**: This involves frequent updates to the application state without triggering user-initiated actions.

- **Animations or rendering optimizations that involve low-level DOM manipulation or canvas drawing operations**: Running the related code outside the Angular zone can enhance performance by bypassing Angular's change detection and allowing for more direct control over rendering updates.

By strategically utilizing `runOutsideAngular()`, you can improve the performance and responsiveness of your Angular application, particularly when dealing with computationally intensive tasks or interactions with external libraries. However, it's crucial to balance performance optimization with maintaining the integrity and functionality of your application. When running tasks inside `runOutsideAngular()`, change detection will not detect the tasks, so you risk showing incorrect data to the user. A good countermeasure to this is running the heavy computation inside the `runOutsideAngular()` method, and then assigning the values to your component properties inside the NgZone again by using the `run()` method, as shown in the following code:

```
@Component({......})
export class ExampleComponent {
  protected readonly ngZone = inject(NgZone);
  performTask(): void {
    this.ngZone.runOutsideAngular(() => {
      console.log(<Task performed outside Angular Zone>);

      // Run inside the runOutsideAngular method again
      this.ngZone.run(() => {
        console.log(<Running inside NgZone again>);
      });
    });
  }
}
```

In the preceding code, you can see how to use the `runOutsideAngular()` and `run()` methods. You inject the NgZone and call the methods on the service provided to you by Angular. Within the callbacks of each method, you can perform any logic you want to perform inside or outside of the NgZone.

Now that you know how to use `runOutsideAngular()` and run code outside the NgZone to improve the performance of your application, let's move on to the next tool Angular offers us to develop more performant applications: the `NgOptimizedImage` directive.

Understanding and using the NgOptimizedImage directive

Another crucial aspect when building performant applications is optimizing your images. The load time of your images is a huge factor in the **Largest Contentful Paint** (**LCP**) of your website, which is one of the three Core Web Vital metrics [the other two Core Web Vital metrics are **First Input Delay** (**FID**) and **Cumulative Layout Shift** (**CLS**)]. The LCP indicates the speed at which the primary content of a webpage loads, specifically measuring the duration from when the user triggers the page load to when the largest image or text block is displayed within the visible area of the browser window. Because images mostly take longer to load compared to text content, how your images are loaded and displayed plays a vital role in the LCP of your applications.

Within the Angular framework, you can improve how images are loaded by using the NgOptimizedImage directive. NgOptimizedImage focuses on prioritizing the loading of the LCP images.

By default, this directive enables lazy loading for non-priority images, conserving bandwidth and improving initial page load times. Additionally, NgOptimizedImage generates a preconnect link tag in the document head, optimizing resource fetching strategies. NgOptimizedImage automatically sets the fetchpriority attribute on the img tag, emphasizing the loading priority of the LCP image. Furthermore, the directive streamlines the process of generating srcset attributes. By using srcset attributes, the browsers request images at the right size for the user's viewport and because of that, no time and resources are wasted downloading an image that's too large.

Besides prioritizing the loading of LCP images, NgOptimizedImage ensures a series of image best practices are applied:

- **Image CDN utilization**: The directive encourages the use of image content delivery network (CDN) URLs, facilitating image optimizations and efficient delivery across global networks.

- **Required width and height**: Setting the width and height attributes for your images when using NgOptimizedImage incorrectly or not setting dimensions results in a warning. By setting width and height properties, you mitigate layout shifts, improve your CLS, and ensure proper rendering.

- **Visual distortion warning**: NgOptimizedImage alerts developers to potential visual distortions in rendered images.

Now that you know why you need the NgOptimizedImage directive, let's see how you can use it. The NgOptimizedImage directive is standalone, so you begin by importing the NgOptimizedImage directive directly into the necessary NgModule or standalone component. Next, you can use NgOptimizedImage by replacing the src attribute on an img tag with ngSrc:

```
<img ngSrc="dog.jpg">
```

As mentioned, you also need to set the width and height properties:

```
<img ngSrc="dog.jpg" width="400" height="200">
```

That is everything you need to use the NgOptimizedImage directive, but there are some additional options you can add. Let's start by exploring the priority attribute. When you mark an image with priority, the following optimizations are applied for you:

- fetchpriority=high
- loading=eager

When you use server-side rendering, it automatically generates a preload link element.

You can mark an image with priority as follows:

```
<img ngSrc="dog.jpg" width="400" height="200 priority ">
```

All LCP images should be marked as priority. If you don't mark an LCP image as priority during development, Angular will log an error.

Besides the `priority` attribute, another useful attribute used with the `NgOptimizedImage` directive is the `fill` attribute, like so:

```
<img ngSrc="dog.jpg" fill>
```

The `fill` attribute is used when you want the image to fill the containing element. A good use case for the `fill` attribute is when you want to use the image as a background image, or when you don't know the exact size of your image and want to fit it inside a container of which you do know the size in relation to the screen size. When using the `fill` attribute, you don't have to set the width and height properties, as Angular will set them for you behind the scenes when the sizes are resolved.

To control how the image will fill the container, you can use the `object-fit` CSS property.

> **More information**
>
> Besides the `priority` and `fill` attributes, when using a third-party service for your images, the `NgOptimizedImage` directive has more cool features such as low-resolution placeholders and custom image loaders. These features are out of scope for this book, but if you want, you can read about them in the official Angular documentation at `https://angular.io/guide/image-directive`.

Now that you know about `NgOptimizedImage` and how you can use it to optimize your image and improve the LCP of your applications, let's dive into the next performance optimization step: using the `trackBy` or `track` function for loops inside the HTML template.

Understanding and using the trackBy and track functions

In Angular applications, rendering large lists or collections of data can sometimes lead to performance issues due to frequent DOM manipulations. To optimize the performance of your Angular application, it's crucial you understand and leverage tools such as the `trackBy` or `track` function.

The `trackBy` function is a feature provided by Angular that improves the performance when rendering lists using the `*ngFor` directives. The `track` function is the counterpart of the Angular control flow syntax. The `trackBy` function is optional, whereas the `track` function is required when using the control flow syntax.

By default, Angular uses object identifiers to track changes in the data provided to an `*ngFor` directive. However, this approach can lead to unnecessary re-renders of DOM elements, particularly when dealing with dynamic data. The `track` and `trackBy` functions allow Angular to efficiently track changes in the collection by providing a unique identifier for each item. This results in fewer DOM manipulations and significantly improves rendering performance, especially when dealing with large datasets.

When you use the `*ngFor` directive, you assign the `trackBy` property to a function and declare the corresponding function inside your component class. The function should return the unique identifier you want to use to track the items inside the list you are rendering:

```
<div *ngFor="let item of items; trackBy: trackById">…</div>
```

In the preceding code, you can see the HTML code using the `*ngFor` directive and defining the `trackBy` property. The `trackBy` property is assigned with a function named `trackById`. This `trackById` function has to be declared inside the component class, like this:

```
trackById(index: number, item: Item) { return item.id }
```

In the preceding example, you use the `id` property from the objects you are rendering with the `*ngFor` directive as the unique identifier (this assumes the objects have an `id` property, otherwise you return another unique property). It's important to note that the `trackBy` function should only be used when the items in the collection have a unique identifier. Using a non-unique identifier or omitting the `trackBy` function altogether can lead to unexpected behavior and performance issues.

When using the control flow syntax to output a list inside your HTML template, the syntax is a bit simplified. Instead of a `trackBy` function, you now use the `track` function and directly provide it with the unique property to check instead of creating a function that returns the unique property, like so:

```
@for (item of items; track item.id) { … }
```

Now that you know why you need to use `trackBy` and `track` functions when rendering lists in your HTML templates, let's explore web workers, the next performance optimization that Angular has at its disposal.

Understanding and using web workers in Angular

Web workers allow you to execute CPU-intensive tasks within a separate thread running in the background, thereby making the primary thread free to update the user interface and run the main threat without any hiccups. Whether it involves intricate tasks such as producing **computer-aided design** (**CAD**) drawings or conducting complex geometric computations, applications can leverage web workers to enhance overall performance significantly.

You add a web worker to your application with an Nx generator. Open the **NX console**, click on **generate**, search for web worker, and select the **@nx/angular – web worker** generator. Next, you need to give your web worker a name and select a project to add the web worker to. If you work without Nx, you can run the following CLI command:

```
ng generate web-worker <location>
```

Running the Nx generator or Angular CLI command will configure your project to use web workers if it isn't configured already. It will also generate a file with your web workers. If you named your web worker heavy-duty, the generated file will be named heavy-duty.worker.ts; when using the Angular CLI, the name of the file will equal the location you provided in the CLI command.

Inside the generated worker file, you will find the initial scaffolded code you need for your web worker. When using Nx, you'll find the following code in the generated file:

```
addEventListener('message', ({ data }) => {
  const response = `worker response to ${data}`;
  postMessage(response);
});
if (typeof Worker !== 'undefined') {
  const worker = new Worker(new URL(<./heavy-duty.worker', import.
meta.url));
  worker.onmessage = ({ data }) => {
    console.log(`page got message ${data}`);
  };
  worker.postMessage(<hello>);
} else { // Fallback for environment. }
```

The addEventListener function will stay in the worker file, and the rest of the code must be located in the component or service where you want to use the web worker. By moving everything but the addEventListner function, you can send messages from the component or service to the web worker. As you can see, in the code that must be moved, there is a fallback for environments where the web worker doesn't work. This is because when using server-side rendering, web workers do not work and you need to have a fallback.

To work with the web worker, you need to send messages to and from the web worker to perform the logic you need to perform. For example, let's say you want to use the web worker when a component is initialized. To achieve this, you add the following code inside the component where you want to use the web worker:

```
@Component({......})
export class FooComponent {
  heavyDutyResult;
  heavyDutyInput = {......};
  constructor() { this.runWebWorker() }
```

```
   runWebWorker () {
     if (typeof Worker !== <undefined>) {
       const worker = new Worker(new URL(<./heavy-duty.worker>, import.
meta.url));
       worker.onmessage = ({ data }) => {
         this.heavyDutyResult = data;
       };
       worker.postMessage(this. heavyDutyInput);
     } else { // Fall back }
   }
}
```

As you can see in the preceding code, you use `worker.postMessage` to send a message to the web worker. This is received inside the event listener of the web worker. When the `postMessage()` function is called in the web worker, it will be received in the `worker.onmessage()` callback function inside the component. Now, you only need to update the web worker file to perform the heavy-duty logic:

```
addEventListener('message', ({ data }) => {
  const response = heavyDutyFunction(data);
  postMessage(response);
});
```

As you can see in the preceding code, we perform some logic – in this example, an imaginary `heavyDutyFunction()` – and send the response back to the component using the `postMessage()` function. Now the circle is complete. You can send some data from the component to the web worker and the web worker will receive this data, perform the heavy-duty logic with the data, and returns the `response` constant to the component class.

Now you know how to use a web worker to create multithreading and run resource-intensive code without blocking your main threat. To wrap up the section, I will mention some other methods you can use to improve the performance of your Angular applications:

- **Lazy loading**: Lazy loading routes help to only load sections of your app that the user actually reaches. We already showcased this in *Chapter 2*, but it's worth mentioning as a performance optimalization.

- **Pre-loading routes**: Using the `preloadingStrategy` on your routes, you can also pre-load routes you anticipate the user will navigate.

- **Caching**: You should cache your API requests and the results of heavy tasks you regularly perform. You can build your own caching system by utilizing `Record` classes, for example. For API requests, I can recommend using `ts-cachable`.

- **Using pure pipes**: We already explained the usage of pipes and what pure pipes are in *Chapter 3*, but they are worth mentioning as a performance optimization.

- **Using canMatch**: Using the `canMatch` route guard combined with lazy-loaded routes prevents you from loading modules and components the user is not allowed to access.

- **Using RxJS effectively**: Running code asynchronously doesn't block your threat and can help to improve the performance of your application.

- **Server-side rendering**: By running the `ng add @angular/ssr` command, you can enable server-side rendering, greatly improving the performance of your application. We will not cover server-side rendering in further detail, but as of Angular 17, you can also include page hydration when using server-side rendering, further enhancing the performance.

- **Virtual scrolling**: Virtual scrolling is a feature in the Angular Material CDK that enables you to effectively render large lists. The virtual scroll will ensure that only items within the viewport are rendered.

You now know how to improve the performance of your Angular applications using `OnPush` and Signals, run code outside the NgZone or create multithreading using web workers, optimize images using the `NgOptimizedImage` directive, and render lists in a performant way by utilizing the `trackBy` and `track` functions. You also learned about other tools and tips to further enhance the performance of your Angular applications. Next, we will learn how you can improve the security of your Angular applications.

Building secure Angular applications

In a world where hacks and exploits are more frequent than ever, you are also responsible for developing secure applications. In this section, we'll delve into the various security risks that Angular applications may face and explore strategies to mitigate them effectively.

When it comes to securing frontend applications, you want to ensure that the users can't reach parts of your application they are not intended to go to and that they can't perform malicious actions that will compromise your application. We will first look at the first scenario and ensure that users can't reach sections of your applications they are not intended to reach.

Setting up route guards

Route guards are used to guard specific routes within your Angular application. They prevent unauthorized users from accessing certain parts of your application. For example, most parts of your application should only be accessible to users who are logged in; other routes might be restricted based on user roles or other factors. Within Angular, there are four different types of route guards:

- **canActivate**: Determines whether the user can activate a specific route.

- **canActivateChild**: Determines whether the user can activate the child routes of a specific route.

- **canDeactivate**: Determines whether a user can deactivate a specific route.
- **canMatch**: Determines whether a user can load a specific route. The difference from `canActivate` is that if the `canMatch` guard fails, the module or standalone component related to the route is not loaded at all. Using `canMatch` offers some performance benefits when combined with lazy-loaded routes.

Since Angular 15, route guards have been implemented using a functional approach; in earlier versions, a class-based approach was used. The class-based approach is currently deprecated, so we will only cover the functional approach. You can declare each guard type you want to use in your route configuration, like this:

```
{
  path: <…',
  loadComponent: () => import(<……'),
  canMatch: [],
  canActivate: [],
}
```

As you can see, you define the guards in the route configuration object. Each guard type is assigned an array containing the guard function that it should resolve before the user can access the route. Each guard function returns a Boolean: `true` if the guard passes and the user can access the route, or `false` if the guard fails and the user can't access the route.

In its simplest form, you can define the guard function directly inside the array assigned to the guard type property:

```
canMatch: [() => inject(UserService).loggedIn],
```

In the preceding example, we `inject` a service and check whether the user is logged in (we did not create the service in this book; this is just an example). If the `loggedIn` property is `true`, the user can access the route. If the `loggedIn` property is `false`, the user can't access the route.

In some scenarios, you might need access to route properties or the current component. If this is the case, you create a function that implements the `CanActivateFn`, `CanActivateChildFn`, `CanDeactivateFn`, and `CanMatchFn` type aliases. When using these type aliases, Angular provides the function with some function parameters you can use inside the guard logic:

- **CanActivateFn**: Contains the function parameters: route of type `ActivatedRouteSnapshot` and state of type `RouterStateSnapshot`.
- **CanActivateChildFn**: Contains the same function parameters as the `CanActivateFn` type alias.

- **CanDeactivateFn**: Contains the function parameters: component with a dynamic type, `currentRoute` of type `ActivatedRouteSnapshot`, `currentState` of type `RouterStateSnapshot`, and `nextState` of type `RouterStateSnapshot`.

- **CanMatchFn**: Contains the function parameters: `route` of type `Route` and `segments` of type `UrlSegment[]`.

You use the type aliases by defining a function that resolves in a `Boolean`. You type the function with the type alias and include the function parameters inside the function brackets. Here is an example implementing the `CanMatchFn` type alias:

```
export const hasRouteSegements: CanMatchFn = (route: Route, segments:
UrlSegment[]) => {
return inject(UserService).loggedIn && segments.length > 1;
};
```

In the preceding example, we check whether the user is logged in and whether there is more than one route segment. To use this guard, you add it to the array of the `canMatch` property inside the route configuration:

```
canMatch: [hasRouteSegements]
```

You can also directly implement the type alias inside the array without defining the function elsewhere:

```
canDeactivate: [(component: UserComponent) => !component.
hasUnsavedChanges]
```

Now you know how to define functional route guards and prevent unauthorized users from accessing routes they aren't allowed to access.

Although this already makes your application more secure, there are other risks when building Angular applications whereby users can perform malicious activities. So, let's outline some attack surfaces and learn how you can mitigate them.

Angular attack surfaces and how to mitigate them

Before delving into +Angular-specific security measures, it's essential to understand the common threats that web applications face. These threats include **cross-site scripting (XSS)**, **cross-site request forgery (CSRF or XSRF)**, injection attacks, and HTTP-level vulnerabilities such as **cross-site script inclusion (XSSI)**.

Angular has some built-in tools to reduce the security risks of these attacks for you and there are some preventive measures you can take yourself when developing your Angular application. Let's start with the most prevalent risk when developing frontend applications: XSS attacks.

Mitigating XSS attacks

In simple terms, you block XSS attacks by preventing malicious code from entering the **Document Object Model** (**DOM**). For example, XSS attacks can run malicious code on your website if they can trick you into injecting a `<script>` tag into the DOM. Other HTML elements that allow code exaction and can be used by attackers include the `<img>` and `<a>` tags. An attacker can use an XSS attack to hijack user sessions, steal sensitive data, or deface websites.

Angular takes a proactive approach to security, treating all values as untrusted by default. This means that when values are inserted into the DOM via template binding or interpolation, Angular automatically sanitizes and escapes untrusted values. This approach significantly reduces the risk of XSS attacks, a prevalent security vulnerability. Even though Angular proactively sanitizes and escapes untrusted values, there are still some actions you can take to make your applications even safer and protect them from security vulnerabilities.

Values inserted into the DOM via template binding or interpolation are automatically sanitized and escaped if the values are not trusted. On the other hand, Angular trusts HTML templates by default, because of which you should treat HTML templates as executable code. Never directly concatenate user input and template syntax because this would enable an attacker to inject harmful code into your application. Here is an example of what you should avoid:

```
<div>{{ data }}</div> + userInput
```

One way to reduce the template risks is by using the default **ahead-of-time** (**AOT**) template compiler when creating production builds. Because the AOT compiler is the default, you don't have to do anything unless you change the default compile settings.

Other possible attack surfaces for an XSS attack are `style`, `innerHTML`, `href`, and `src` bindings where the bound value is provided by the user:

```
<div [innerHTML]="htmlSnippet"></div>
<div [style]="userProvidedStyles">...</div>
```

Attackers can use unsafe binding to inject harmful code or URLs into your application. Besides unsafe bindings, you also should avoid direct interaction with the DOM. If you bind an unsafe value, Angular will recognize it in most cases and sanitize it by removing the unsafe value. It's good to be aware of this because it can lead to broken functionality in your application. Also, some attackers might be able to circumvent the sanitation, so be careful when using unsafe binding options. If you want to bind a URL, script, or other value that Angular will sanitize and you know the value is safe, you can bypass the sanitation using the `DomSanitizer` service provided by Angular.

If you want to bypass sanitation, you start by injecting the `DomSanitizer` service:

```
protected readonly sanitizer = inject(DomSanitizer);
```

Next, you can use the bypass methods exposed by the service to bypass sanitation:

```
this.trustedUrl = this.sanitizer.bypassSecurityTrustUrl(this.
dangerousUrl);
```

The `DomSanitizer` service exposes five different options to bypass sanitation:

- `bypassSecurityTrustHtml`
- `bypassSecurityTrustScript`
- `bypassSecurityTrustStyle`
- `bypassSecurityTrustUrl`
- `bypassSecurityTrustResourceUrl`

Depending on what value you are bypassing, you use the corresponding `bypassSecurity` method, so to bypass the sanitation of a piece of HTML, you would use the `bypassSecurityTrustHtml` method.

Besides binding unsafe values, another possible attack surface is direct manipulation of the DOM. The built-in browser DOM APIs don't protect you from security vulnerabilities unless `Trusted Types` are configured. For example, elements accessed through `ElementRef` instances, the browser document, and many third-party APIs contain unsafe methods. You should avoid interacting with the DOM directly and instead use the `Renderer2` service when you need to manipulate DOM nodes.

Lastly, you can configure a **Content Security Policy** (**CSP**) to prevent XSS attacks. A CSP can be enabled on the web server and falls out of scope for this book.

You now know what XSS attacks are, what Angular does to prevent them, and what measures you can take to prevent them. Next, you will learn what vulnerabilities there are when making HTTP requests and what you can do in your Angular applications to prevent them.

Mitigating HTTP-related security risks

Ensuring robust security measures against HTTP-related risks is paramount to safeguarding your application and its users. Two significant threats to consider are CSRF (or XSRF) and XSSI. In this section, we will dive deeper into CSRF and XSSI and explain what they are, how they can affect your applications and users, and what measures you can take to prevent CSRF and XSSI exploits.

While CSRF and XSSI predominantly have to be mitigated on the server side, Angular does provide some tools to make the integration with the client side a bit easier. We will start by explaining what CSRF is and what you need to do on the client side to prevent it.

What CSRF/XSRF attacks are and how to prevent them

Imagine you're logged into your online banking account in one tab of your browser. Now, if you visit a malicious website in another tab, that site can secretly make requests to your banking website without your knowledge. These requests could transfer money, change your password, or perform any action that your banking website allows – all without your consent.

CSRF/XSRF attacks can have serious consequences. They can lead to unauthorized transactions, data manipulation, and even account takeovers. Since the attacker doesn't need to know your login credentials, these attacks can bypass traditional authentication mechanisms.

To protect against CSRF/XSRF attacks, websites typically use techniques such as CSRF tokens. These tokens are unique identifiers generated by the server and sent to the frontend. The frontend includes these random tokens with each request so the server can verify the token, ensuring that the request originated from a legitimate source and not from a malicious website. Commonly, the token is sent to the frontend using a cookie flagged with `SameSite`. If the cookie also includes the `httpOnly` flag, you don't have to do anything on the frontend and everything will be handled on the backend, but this isn't always the case; often, you must include the token in the request headers.

Using a CSRF token is an effective measure because all browsers have the same-origin policy. The same-origin policy ensures that only the code of the website where a cookie is set can read the cookie. The same-origin policy also ensures that a custom request header can be set by the code of the application making the request. That means that malicious code from the website the attacker tricked you into using cannot read the cookie or set the headers for your request. Only the code of your own application can do this.

If the cookie with the CSFR token is not an `httpOnly` cookie and the client is required to add the cookie in the request header, you can create an HTTP interceptor for this purpose. Here is an example of how the interceptor could be implemented:

```
export const MockInterceptor: HttpInterceptorFn = (
  req: HttpRequest<unknown>,
  next: HttpHandlerFn,
) => {

  const csrfToken = inject(AuthService).getCsrfToken();
  const csrfReq = req.clone({
    setHeaders: {
      <X-XSRF-TOKEN>: csrfToken,
    },
  });

  return next(csrfReq);
};
```

Besides adding a CSRF token, there isn't anything you can do on the frontend to protect your application from CSRF attacks. If you need to add the token, it depends on how the server side implements the cookie, so consult with the backend team about this topic.

Now that you know what CSRF/XSRF attacks are, let's learn about XSSI attacks.

What XSSI attacks are and how to prevent them

XSSI attacks occur when an attacker injects malicious scripts into a web page from an external domain. These scripts are executed in the context of the victim's session, potentially compromising sensitive information and performing unauthorized actions. XSSI attacks can lead to data theft, session hijacking, and unauthorized manipulation of user interactions.

XSSI attacks are also known as the **JSON vulnerability** because this vulnerability exploits weaknesses in older browser versions by manipulating native JavaScript object constructors. By overriding these constructors and embedding an API URL within a `<script>` tag, malicious actors can execute unauthorized requests and retrieve sensitive information from the targeted JSON API.

The success of this exploit hinges on the JSON data being executable as JavaScript. To prevent XSSI attacks, servers can adopt a preventive measure by prefixing all JSON responses, rendering them non-executable. Conventionally, this is achieved by appending the widely recognized)] } ' , \n string.

The `HttpClient` of the Angular framework is equipped to handle this security measure seamlessly. It detects and removes the)] } ' , \n string from incoming responses automatically before proceeding with further parsing, thus fortifying the application against potential exploits. Because Angular automatically detects the)] } ' , \n string and removes it for you, you don't have to do anything for XSSI prevention in the frontend, but it's always good to be aware of the attack and how it actually can be prevented. If your backend team uses a different prevention measure, align with it to see whether you need to do anything in the frontend.

Summary

In this chapter, you learned about performance and security. You took a deep dive into Angular change detection, giving you a better understanding of how Angular detects changes and how you can reduce the number of components and bindings that Angular has to check when performing change detection.

You also learned about other measures you can take to ensure your Angular applications remain performant. You learned how to run code outside of the Angular zone, you learned about the `NgOptimizedImage` directive, you learned about the `trackBy` and `track` functions, and you've created your own web worker to run code in a separate threat. Furthermore, you learned that you can use lazy loading, `canMatch`, server-side rendering, and other tools provided by the Angular framework to enhance application performance even more.

After taking a deep dive into Angular application performance, you learned how you can develop secure frontend applications using the Angular framework. You learned how to prevent users from accessing pages they aren't intended to reach. You also learned about common exploits, what measures Angular takes to prevent these attacks, and what steps you can take to make your application even more secure.

In the next chapter, you will learn how to make your applications more accessible and tailored to the users visiting them. You will learn about translatable content, using the correct formatting and symbols for each user, and making your website accessible to people of all abilities.

10

Internationalization, Localization, and Accessibility of Angular Applications

When developing applications, you're often targeting users from many different countries; because of that, it's essential to add internationalization and localization to your Angular applications. **Internationalization** is the process of making your applications translatable so people who speak different languages can use your application without any issues. With **localization**, you tailor the content of your website to a specific location. For example, users from the USA expect dollar signs, whereas users from the EU use a euro sign in front of currency values. Besides people from different countries, your applications will also be used by people with different abilities. Some users might not be able to use a keyboard or read the screen like other users. In a world where we rely more on applications, it's important to develop your applications in a way that ensures users of all abilities can use them.

In this chapter, you will learn how to develop Angular applications that can serve as many people as possible. First, we will dive into and implement internationalization within the Nx monorepo. After making your content translatable, you will learn about and implement localization to correctly display your dates, currencies, and other values depending on the user. Lastly, we will dive into the topic of developing accessible frontend applications that can be used by users of all abilities.

This chapter will cover the following topics:

- Adding translatable content in Angular applications
- Localization for Angular applications
- Making your Angular applications accessible to everyone

Adding translatable content in Angular applications

In this section, we will dive into internationalization, commonly referred to as **i18n** (i18n is the abbreviation for internationalization, where "i" and "n" are the first and last letters of the word, and 18 stands for the number of letters in between the "i" and the "n"). Internationalization is developing usable applications for people who speak different languages. Simply put, when you implement i18n, you make your application content translatable. About 75% of the world doesn't speak English at all, so if you only display your application content in English (or exclusively use another language), you are missing out on a lot of potential users.

The Angular framework has built-in functionalities to support translatable content, but we will not be using the built-in i18n solution to support translatable content. The first reason is that the built-in solution used XML content in the translation files. Most developers don't like XML, and JSON is a much more readable and flexible alternative. The second reason we will not use the built-in i18n solution for translatable content is that it's predominately focused on compiling time translations. With compile-time translations, the translation keys are replaced during the compilation process; because of that, you must create and deploy a specific build for each language you want to support. There is also some support for runtime translations, meaning the translations can change when the application is deployed and running, but there is little documentation, and the solution is rarely used because there are much better options developed and maintained by the Angular community.

The two most commonly used i18n libraries that support translatable content in Angular applications are **Transloco** and **ngx-translate**. We will use Transloco, as it has better support for standalone components, is more actively maintained, has better documentation, supports server-side rendering, and has more configuration options and features overall. We will implement Transloco step by step into the *expenses-registration application* in the Nx monorepo.

You can still dive into Transloco's documentation at this website: `https://jsverse.github.io/transloco/`.

Installing Transloco in your Nx monorepo

Before you can support translatable content in your HTML templates and TypeScript files, you need to install Transloco in your Nx monorepo and add it to the applications where you want to use it. So, let's start at the beginning and install the NPM package using the following command:

```
npm i @ngneat/transloco
```

After installing the package, you need to add Transloco to your project, open the terminal at the `business-tools-monorepo\apps\finance\expenses-registration` folder path, and run the following command in your terminal:

```
nx g @ngneat/transloco:ng-add
```

When you run this command, you will be asked two questions inside the terminal:

- **Which languages do you need?**: Provide comma-separated values for the languages you want to add. I used en to add English and nl to add Dutch as my default languages. You can provide your own languages, but please add en for English, as we will use that one in the book. You can always extend upon these languages if you want.

- **Are you working with server-side rendering?**: False–this is the default, so you can just hit *Enter*).

After answering the questions, the command will add the necessary configuration to add Transloco to the *expenses-registration application*. You can also add the configurations manually, but using the terminal command is faster and ensures that everything is added correctly.

Now let's look closer at the files Transloco generated for you that have been added to the Nx monorepo and *expenses-registration application*.

The first file we will look at is `transloco-loader.ts`. Inside this file, Transloco created an Angular service that is responsible for loading the translation files based on a provided language. The following code is generated for you in the `transloco-loader.ts` file:

```
@Injectable({ providedIn: 'root' })
export class TranslocoHttpLoader implements TranslocoLoader {
    private http = inject(HttpClient);
    getTranslation(lang: string) {
        return this.http.get<Translation>(`/assets/i18n/${lang}.
json`);
    }
}
```

As you can see in the preceding code, the `TranslocoHttpLoader` service does fetch a JSON file from your assets folder. The fetched JSON file will contain translation keys and values for a specific language. You need to provide the `getTranslation()` function with a language key that correlates with the naming of your translation file. For example, if you have an `en.json` file inside the `i18n` folder containing your English translations, you would call the `getTranslation()` function with en as a parameter to fetch the English translations.

Personally, I like to ensure better typing for the function parameter of the `getTranslation()` function and only allow specific string values instead of any string. In my case, I have en and nl as my language files, so I would adjust the function parameter to this:

```
lang: 'en' | 'nl'
```

You can adjust the function parameter type to the languages you support.

> **Important note**
>
> It's important to note that when you deploy your Angular application and Transloco isn't able to load your language files, it might be because you need to provide a relative path, but generally speaking, this is not the case.

Now let's see how you can use the `getTranslation()` method with a relative path:

```
getTranslation(langPath:'en' | 'nl') {
    return this.http.get(`./assets/i18n/${langPath}.json`);
}
```

In the preceding code, you'll find an example of a relative path, but you only have to provide a relative path if Transloco can't load your translations with the regular setup.

Now that you know about the `transloco-loader.ts` file and what it is used for, let's look at what else Transloco created for you.

For each language you provided when you ran the command in the terminal, Transloco created a translation file for you. Inside your assets folder, an `i18n` folder was created, and inside the `i18n` folder, you'll find a JSON file for each of the languages you provided. So, in my case, I provided `en` and `nl`, so Transloco created an `en.json` and `nl.json` file for me. You will add your translation keys and values to these JSON files. By default, the files contain an empty JSON object. After generating the empty translation files, Transloco adjusted the `ApplicationConfig` object inside the `app.config.ts` file.

Inside your `ApplicationConfig`, Transloco added the `provideHttpClient()` and `provideTransloco()` configurations inside the providers array:

```
provideHttpClient(),
provideTransloco({
    config: {
        availableLangs: [<en>, <nl>],
        defaultLang: <en>,
        // Remove this option if your application doesn>t support changing
language in runtime.
        reRenderOnLangChange: true,
        prodMode: !isDevMode(),
    },
    loader: TranslocoHttpLoader
}),
```

The `provideTransloco()` configuration function receives a configuration object of its own, configuring the Transloco instance. By default, the configuration for Transloco defines the translation loader, the default language, the available languages, whether you are in production mode, and whether you support runtime language changes within your application. Additionally, you can add other configurations such as a fallback language, the number of times Transloco needs to try loading a language file before using the fallback language, or if it should log missing translation keys.

Remove the generated provideHttpClient() function

Transloco also added the `provideHttpClient()` function inside your providers array. Because you already have a `provideHttpClient()` configuration inside the providers array in your `ApplicationConfig` object, it's important to remove this configuration added by Transloco. If you do not remove the `provideHttpClient()` added by Transloco, it will overwrite your own `provideHttpClient()` configuration, which also contains your mock data interceptor configuration.

You need to adjust one more thing because of the mock data interceptor. Go to the `mock.interceptor.ts` file and adjust the first `if` check. We want to return the request without modifying it in production (as we already do) or if the request URL ends with `.json`:

```
if (!isDevMode() || req.url.endsWith('.json')) return next(req);
```

If you now check the network tab of your browser console, you'll see that a request is made and successfully fetches the `en.json` file.

Now that you know what Transloco generated and added to your application, let's move on and learn how you can use Transloco to support content in multiple languages.

Translating content using Transloco

To translate values using Transloco, you create key-value pairs in your translation files. You use the same key in each translation file and provide it with the correct translation value for the given language file. The keys are all lowercase, and words are separated by an underscore. For example, if I wanted to translate `Expenses Overview`, I would add the following in the `en.json` translation file:

```
{
  "expenses_overview": "Expenses Overview"
}
```

In the other translation files, you add the same `expenses_overview` key and assign it the translation value for that specific language. For example, in my `nl.json` file, I would add the same key with the Dutch translation:

```
{
    "expenses_overview": "Uitgavenoverzicht"
}
```

Now that you know how to add translation keys and values to the translation files, let's see how to use these key-value pairs inside your application to translate content. You can translate values inside your HTML templates and the TypeScript files. We'll start by looking into the HTML translations, and you will learn about translating values inside the TypeScript files after that.

Translating values in the HTML template

Inside HTML templates, you can translate values with a **structural directive**, an **attribute directive**, or a **translation pipe**. It's recommended to use the structural directive. Both the pipe and directives create subscriptions to observe when the user changes the selected languages. When using the structural directive, you only create a single subscription for the template. In contrast, the pipe and attribute directives create a subscription each time you use them in a template, which can be multiple times. Also, the structural directive caches the translations, so if you're using the same translation key numerous times in a template, the structural directive can return it directly from the cache, improving your performance and memory usage.

Because the structural directive is the recommended approach, let's first learn how to use that before we look at the pipe and attribute directive. Because we use a standalone component, we need to import the directive in the imports array of the component before we can use it:

```
imports: [......, TranslocoDirective],
```

After importing the directive, add it to the expenses overview page. When using the structural directive, it's recommended to wrap the entire template in an `ng-container` element and add the structural directive on the `ng-container` element:

```
<ng-container *transloco="let t"> ...... </ng-container>
```

When you wrap your template in the `ng-container` element and add the directive on `ng-container`, you can use the `t` variable from the directive throughout the template to translate values. Let's replace `Expenses Overview` text inside the `h1` tag with the translatable value using the `t` variable from the directive and the `expenses_overview` translation key:

```
<h1>{{ t('expenses_overview') }}</h1>
```

As you can see in the preceding code, you can use the t variable from the directive as a function and provide it with the translation key you want to display. In the browser, you'll see **Expenses Overview**, just like before. Only now is it using the translation key value to display the value based on the set language; that's everything you have to do for simple translations. Add the translation key-value pair to the translation files and translate it inside your templates using the t function provided by the *transloco structural directive. But there are also many scenarios where you need to translate something and need dynamic values inside the translated value.

For example, let's say you want to change the title inside the h1 tag from Expenses Overview to Expenses overview for <user name>. In that case, you need a way to provide the username to the translatable value so it can be inserted in the correct place. You also can't just append the username because the build-up of your translatable value might not be the same for each language, or your dynamic value has to be inserted somewhere in the middle of your translated value. Luckily, there is a simple solution for this—you can add parameters inside your translation values and provide a value for these parameters whenever you use the translation key connected to the translation value.

To achieve this, start by adding another (or changing the existing) translation key inside the translation files:

```
"expenses_overview_for_user": "Expenses overview for {{user}}"
```

As you can see, we've added a parameter in our translation value by using double curly brackets. Now, inside the HTML template, you can provide a value for the user translation parameter like this:

```
<h1>{{ t('expenses_overview_for_user', { user: user.fullName }) }}</h1>
```

As you can see, you can provide an object to the t function as a second parameter. The first parameter for the t function is the translation key and the second parameter is optional and can be used to provide values for the translation parameters. In our example, we use a user object with a fullName property to provide the user name to the translation parameter (we currently don't have this user object in our component; it's just an example, and you can provide the user translation parameter with any property or static value you want). Besides parameters and simple translation keys, you can also group translation keys inside the translation files.

To group translation keys, create a grouping key, for example, expenses_overview_page, and add nested keys under the group key like this:

```
{
    "expenses_overview_page": {
        "title": "Expenses Overview",
        "incl_vat": "Incl. VAT"
    }
}
```

Especially when your translation files grow larger, this helps to locate specific translations quickly. Using this approach, you might end up with some duplicate translation values. Still, the improved maintainability is a more significant win than having a few duplicate translation values, but your approach is up to you and your team. When using the grouped approach, you use the following syntax inside your HTML template:

```
<h1>{{ t('expenses_overview_page.title') }}</h1>
```

As you can see in the preceding example, you first define the group name followed by a dot and then the translation key. As you might imagine, prefixing every translation key with the group name might get redundant, so to make things easier, you can define this group name inside the *transloco directive:

```
*transloco="let t; read: 'expenses_overview_page'"
```

When you define the group name inside the directive, you only have to use the inner translation keys inside the t functions:

```
<h1>{{ t('title') }}</h1>
```

You now know how to declare translation key-value pairs in groups or regularly and how to translate them inside your HTML templates using the *transloco structural directive. Alternatively, you can use the transloco attribute directive or pipe to do your translations; however, as mentioned, the attribute directive is recommended because it creates fewer subscriptions and has caching throughout your HTML template.

To give a full overview of the possibilities, here is an example including translation parameters using the pipe:

```
<h1>{{ 'expenses_overview_for_user' | transloco: {user: user.fullname}
}}</h1>
```

Lastly, you can use the attribute directive:

```
<h1 transloco=" expenses_overview_for_user " [translocoParams]="{
user: user.fullname }"></h1>
```

In the preceding example, we first showcased how you can use the transloco pipe to translate translation keys. In the last example, we showed how you can use the transloco attribute directive combined with the translocoParams attribute directive to translate translation keys with parameters. But, as mentioned earlier, using the structural directive is the preferred approach, as it improves your performance and creates fewer subscriptions.

Now you know about all the options to translate values inside your HTML templates using Transloco. Next, you'll learn about translating values programmatically inside your TypeScript files.

Translating values programmatically in your TypeScript files

Sometimes, you might encounter a situation where you must translate values inside your component classes or services. To translate values programmatically inside your TypeScript files, you can use the `TranslocoService` class. There is only one caveat: you need to know the translation file is loaded before you can translate translation keys inside your TypeScript files.

So, let's start by creating a `TranslationService` class in the *expenses-registration application*. Inside the `TranslationService`, we will make the `TranslocoService` class publicly available and create a signal from the `events$` observable using the `toSignal()` function:

```
@Injectable({ providedIn: 'root' })
export class TranslationService {
  translocoService = inject(TranslocoService);
  translationsLoaded = toSignal<boolean>(this.translocoService.
events$.pipe(filter(event => event.type === <translationLoadSuccess>),
map(event => !!event)));
}
```

We injected the `TranslocoService` inside `TranslationService`. Alternatively, you can inject the `TranslocoService` into each component or service you need to, but I prefer to expose it in the `TranslationService`, where we also declare the `translationsLoaded` signal and possible other configurations and methods. The `translationsLoaded` signal you created will return `true` when the `events$` observable from the `TranslocoService` returns an event of type `translationLoadSuccess`. You can use this signal in combination with a signal effect to translate values inside your TypeScript files safely.

For example, if you want to translate the `expenses_overview_page.title` key inside the `ExpensesOverviewPageComponent`, you start by injecting the `TranslationService` service you created:

```
protected readonly translationService = inject(TranslationService);
```

After injecting the service, you can create an effect based on the `translationsLoaded` signal and perform your translations safely inside the effect:

```
translationEventsEffect = effect(() => {
  if (this.translationService.translationsLoaded()) {
    // Perform your translations here
  }
});
```

As you can see in the preceding code, we declare an `effect` and check if the `translationsLoaded` signal from the service returns a `true` value; if it does, you can perform your translations safely. If you know the translation file is already loaded, for example, because the user is logged in and you fetch the translation file before the user is logged in, you don't need the effect and can directly get the translation. You translate a value by calling the `translate()` function on the `TranslocoService` and providing the function with the translation key you want to translate:

```
this.translationService.translocoService.translate('expenses_overview_
page.title')
```

Using the `translate()` function will return the translated value synchronously, so you can directly use or assign it however you need to. When you need to translate a key-value pair that includes translation parameters, you use the following syntax:

```
this.translationService.translocoService.translate(' expenses_
overview_for_user', {user: user.fullname})
```

Alternatively to the `translate()` function, you can also use the `selectTranslate()` function. The `selectTranslate()` function is asynchronous and will fetch the translation file (unless it's already loaded) and return the translation as an Observable. When using the `selectTranslate()` function, you don't have to make sure the translation file is already loaded, but you do need to manage the subscription. I like to use the first approach, as you don't have to create and manage subscriptions each time you want to translate a value inside your component classes, but here is an example of how you can get the translation value for a key using the `selectTranslate()` function:

```
this.translationService.translocoService.selectTranslate('expenses_
overview_page.title').subscribe((title) => {
  console.log(<==>>, title);
});
```

You now know how to translate values synchronously and asynchronously in your TypeScript files. You also know how to translate values in the HTML template. The last thing to learn is how to change your Transloco instance's configurations at runtime. After all, you want to give the user the ability to change the language.

Changing the Transloco configurations at runtime

The last step in making your content translatable for your users is adding the option to change the language. First, you need something so the user can select a language; I will add a select box to the navbar for this purpose. In the HTML file of the `NavbarComponent`, I've added the following HTML:

```
<div>
  <select #selectList (change)="languageChange.emit(selectList.
value)">
    <option *ngFor="let lang of languages()" (click)="languageChange.
emit(lang)" [value]="lang">{{ lang }}</option>
```

```
    </select>
  </div>
```

After adding the select box in the HTML file, in the component class, I've added an input for the languages array and an output to output the selected value (I also converted the navbarItems input to a signal input):

```
export class NavbarComponent {
  navbarItems = input([], { transform: addHome });
  languages = input<string[]>([]);
  @Output() languageChange = new EventEmitter();
}
```

After updating the NavbarComponent, you need to provide an array of languages to the NavbarComponent inside the HTML template of the AppComponent of your expenses-registration application. To achieve this, we're first going to add an additional method in our LanguageService to retrieve the available languages as a string array:

```
getLanguages() {
  return this.translocoService.getAvailableLangs() as string[];
}
```

After adding the method that returns the available languages, we will use this method inside the HTML template of our AppComponent and use it to set the language input for the NavbarComponent:

```
<bt-libs-navbar ...... [languages]="translationService.getLanguages()" />
```

After supplying the languages input with the available languages, you'll see the options inside the select box. Next, you need to add the languageChange output event on the HTML tag of the NavbarComponent and use the output value to set a new active language:

```
<bt-libs-navbar ...... (languageChange)="translationService.
translocoService.setActiveLang($event)" />
```

Now, if you change the language in the select box inside your navbar, you'll notice that the translatable text in your application has been changed to the selected language. That is all you need to know about adding translations in your Angular applications using Transloco. The library has other translation options, such as setting the default language, setting the fallback language, or subscribing to language changes to perform logic reactively when the active language changes.

If you want, you can dive deeper on your own by visiting the documentation on the official website: https://jsverse.github.io/transloco/.

You now know how to support internationalization for your Angular websites using Transloco. We did all the necessary setup for Transloco and created an additional `TranslationService`. Next, we created translation files with key-value pairs used to declare all your translations. You learned how to use the Transloco directives and pipe to translate values inside your HTML templates and the `TranslocoService` to translate values and change the active language inside your TypeScript files. Now that you know how to add i18n support to your Angular applications, let's learn about localization.

Localization for Angular applications

Where internationalization ensures your application content can be consumed in different languages, localization ensures that your application uses the correct formats and symbols for the user's location. For example, dates are formatted differently for users in the USA and in the EU. You want to ensure that you show the correct formats for each user so your application can be used without confusion by as many people as possible. Where internationalization is commonly referred to as i18n, localization is commonly referred to as **l10n**.

Just as with i18n, Angular has its own localization packages, but the implementation of the framework isn't widely used and requires some additional work to use at runtime. Just as we did with i18n, we can use Transloco for l10n. Transloco provides a dedicated package for l10n containing a date, currency, and decimal pipe. To start using the Transloco l10n pipes, start by installing the NPM package with the following command:

```
npm i @ngneat/transloco-locale
```

After installing the package in your Nx monorepo, you need to provide configurations in the `ApplicationConfig` object of the applications you want to use the l10n package. We will add it to the *expenses-registration application* by adding the following provider inside the providers array of the `ApplicationConfig` object:

```
provideTranslocoLocale({
   langToLocaleMapping: { en: <en-US>, nl: <nl-NL> }
})
```

As you can see, you need to add the `provideTranslocoLocale()` provider function inside the providers array and supply the function with a parameter to configure the language to localization mapping. In the example, we map the en language to the `en-US` localization and the nl language to the `nl-NL` localization. By using the `langToLocaleMapping` configuration, the localization will automatically change when you set a new active language. If you don't use the `langToLocaleMapping` setting and want your localization settings separated, you need to set the active location using the `TranslocoService`:

```
this.translocoService.setLocale('en-US');
```

Besides the `langToLocaleMapping` configuration, you can provide the following configurations to the `provideTranslocoLocale()` provider function:

```
export interface TranslocoLocaleConfig {
    defaultLocale?: Locale;
    defaultCurrency?: Currency;
    localeConfig?: LocaleConfig;
    langToLocaleMapping?: LangToLocaleMapping;
    localeToCurrencyMapping?: LocaleToCurrencyMapping;
}
```

After configuring the `provideTranslocoLocale()` provider function, you can start using the Transloco i10n pipes. Let's start with localizing currencies.

Localizing currencies using the translocoCurrency pipe

When localizing currency values inside your applications, you want to display the correct currency symbol and make sure your numbers are formatted correctly depending on the user. To localize currency values with Transloco, you import the pipe into your component (if you're using standalone components) and change the Angular currency pipe inside your templates for the `translocoCurrency` pipe. So go ahead and change the currency pipe in your `expenses-overview-page.component.html` file and replace it with the transloco pipe:

```
<td>{{ expense.amount.value.toFixed(2) | translocoCurrency }}</td>
```

After changing the pipe in your HTML template, your currency symbol will change from a USD symbol to a euro symbol when you change the language from `en` to `nl` inside the navbar (in a production environment, the backend should also return different data because of the exchange rate between the currencies, or the API can provide the exchange rate so you can perform the conversions on the frontend).

You can provide the `translocoCurrency` pipe with some additional options for further configuring the displayed value:

- `display`: This controls what you want to display for the currency unit. The options you can choose from are code, symbol, narrowSymbol, and name.

- `numberFormatOptions`: This is an object that controls how the numbers are formatted. You can provide the object with `Intl.NumberFormatOptions` properties (`Intl.NumberFormatOptions` are native JavaScript format options).

- `currencyCode`: With this option, you can specify the currency symbol that should be used by the pipe.

- `locale`: With this option, you can provide a locale option, such as en-US or nl-NL.

The following is an example of the `translocoCurrency` pipe including parameters:

```
translocoCurrency: 'narrowSymbol' : {minimumFractionDigits: 2 } :
'EUR' : 'nl-NL'
```

The configuration options are optional and will overwrite any default settings you configured inside the `provideTranslocoLocale()` function.

Now that you know about the currency pipe, explore the other localization pipes Transloco has to offer.

Localizing dates using the translocoDate pipe

When localizing dates, you want to ensure the dates are formatted correctly depending on the language settings (or localization settings if you have them separated) the user has selected. Dates in the EU and USA, for example, are formatted differently. For instance, in the USA, dates start with the month. In the EU, they begin with the day. So, if you want to display the 20th of January, 2024 for users from the USA, that would be 1/20/2024, and for EU users, 20-01-2024 would be the conventional formatting. Also, for English-speaking users, months are always capitalized, whereas this is not the case for some EU languages.

To provide the best possible user experience, dates in your applications should be formatted correctly. The `translocoDate` pipe offers a simple way to achieve this. Just import the pipe into your component (if you're using standalone components). Similar to the currency pipe, you can replace the native Angular date pipe with the `translocoDate` pipe, and your dates will automatically react when the user changes the language (or localization) settings. To demonstrate this, you can change the date pipe you used in the `expenses-overview-page.component.html` file:

```
{{ expense.date | translocoDate }}
```

After changing the pipe to the `translocoDate` pipe, the date format changes when you change the language settings inside the navbar. Additionally, you can provide the `translocoDate` pipe with some parameters to further configure how formats are displayed or to overwrite the default settings for specific instances in your application. The `translocoDate` pipe can take the following parameters:

- `options`: This is an object containing the following properties:

  - `dateStyle`: Controls how to format the date; you can use full, long, medium, and short formats.

  - `timeStyle`: Controls how to format the time of the date. You can provide it with full, long, medium, and short.

  - `fractionalSecondDigits`: Controls the number of fractional seconds to show; you can set it to 0, 1, 2, or 3.

- `dayPeriod`: Controls how to show the period of the day (at night, midday, etc.). The options are long, short, and narrow.

- `hour12`: Indicates whether you are using 12 or 24 hours. You can provide the `hour12` property with a `true` or `false` value.

- `weekday`: Controls how weeks are formatted; the options are long, short, and narrow.

- `era`: Controls how the era is formatted; the options are long, short, and narrow.

- `year`: Controls how to format the year value; the options are numeric and 2-digit.

- `month`: Controls how to format the month value; the options are numeric, 2-digit, long, short, and narrow.

- `day`: Controls how to format the day value; the options are numeric and 2-digit.

- `hour`: Controls how to format the hour value; the options are numeric and 2-digit.

- `minute`: Controls how to format the minute value; the options are numeric and 2-digit.

- `second`: Controls how to format the second value; the options are numeric and 2-digit.

- `timezone`: Controls how to display the time zone; the options are full, long, medium, and short.

- `locale`: With this option, you can provide a locale option, such as en-US or nl-NL.

All the additional parameters you can supply to the `translocoDate` pipe are optional; if you don't supply any additional parameters, the pipe will display dates with the numeric day, month, and year by default. Now that you know how to use the `translocoDate` pipe and what configuration options it has, let's move on to the last Transloco pipe we will cover: the decimal pipe.

Localizing numbers using the translocoDecimal pipe

Just like currency and date values, numbers are subject to localization. For example, in the USA, large numbers are separated by a comma and decimal numbers are separated from the integer values by a dot; for EU users, this is reversed. This localization issue can be solved in your applications by using the `translocoDecimal` pipe. It will format numbers correctly depending on the language (or localization) settings configured by the user:

```
<!--1,234,567,890  en-US -->
<span> {{ 1234567890 | translocoDecimal }}</span>
```

As shown in the preceding code sample, when the localization is set to en-US, the large number is separated by commas. If you make it a decimal number, the decimal values will be separated by a dot:

```
<!--567,890.15  en-US -->
<span> {{ 567890.15 | translocoDecimal }}</span>
```

Just as with the `translocoDate` and `translocoCurrency` pipes, you have some additional configuration options for the `translocoDecimal` pipe. The `translocoDecimal` pipe can take the following parameters:

- `numberFormatOptions`: This is an object containing the `Intl.NumberFormatOptions` formatting properties (`Intl.NumberFormatOptions` is a native JavaScript formatting object)

- `locale`: With this option, you can provide a locale option, such as en-US or nl-NL

You've now learned how you can use the different localization pipes of the Transloco library and what configuration options you can provide to the pipes to control the output. In the next section, you'll learn how to make your Angular applications accessible to everybody.

Making your Angular applications accessible to everyone

In a world that relies more on web applications, ensuring everyone can use your application is essential. Making your Angular application accessible to people with motor or visual impairments is crucial in ensuring that users of all abilities can effectively interact with it. **Accessibility**, often abbreviated as **a11y**, involves designing and developing your application to be usable by people with diverse needs, including those with disabilities.

As a developer, I often forget that not everyone can use a keyboard or see a screen just like I can. Not only people with permanent disabilities but also people who temporarily can't use their hands or sight should be able to keep interacting with your applications. In some countries, it's even enforced by law that your application has to implement specific accessibility standards. A commonly used standard for accessibility is the **Web Content Accessibility Guidelines 2.2 (WCAG 2.2)**. WCAG 2.2 has 13 guidelines to follow:

1. **Text alternatives**: Provide text alternatives for non-text content to make it accessible to users who cannot see images or hear audio.

2. **Time-based media**: Provide alternatives for time-based media, such as audio and video, to ensure accessibility for users who cannot hear or see multimedia content.

3. **Adaptable**: Create content that can be presented in different ways without losing information or structure, ensuring accessibility for users who rely on various assistive technologies.

4. **Distinguishable**: Ensure sufficient contrast, provide alternatives for audio content, and avoid distractions that could hinder accessibility to make it easier for users to see and hear content.

5. **Keyboard accessible**: Make all functionality available from a keyboard interface, ensuring accessibility for users who cannot use a mouse.

6. **Enough time**: Provide users with enough time to read and use content, ensuring accessibility for users who may need more time to interact with web pages.

7. **Seizures and physical reactions**: Do not design content in a way that is known to cause seizures or physical reactions, ensuring accessibility for users with photosensitive epilepsy or other conditions.

8. **Navigable**: Make web pages navigable and predictable, ensuring accessibility for users who rely on navigation aids or have cognitive disabilities.

9. **Input modalities**: Ensure compatibility with different input modalities, such as touch, speech, and gestures, ensuring accessibility for users with disabilities that affect how they interact with web content.

10. **Device independence**: Ensure compatibility with various devices, platforms, and assistive technologies, ensuring accessibility for users who use different devices to access the web.

11. **Readable**: Make text content readable and understandable, ensuring accessibility for users with cognitive disabilities or those who use screen readers.

12. **Predictable**: Make web pages operate in predictable ways, ensuring accessibility for users who rely on consistent navigation and interaction patterns.

13. **Input assistance**: Help users avoid and correct mistakes, ensuring accessibility for users who may have difficulty entering information or navigating web forms.

Now that you know the 13 guidelines of WCAG 2.2, let's examine what you can do in your Angular applications to adhere to these guidelines.

How to make Angular applications accessible

The easiest way to adhere to accessibility guidelines is by using a UI library that takes these concerns out of your hands. Some good Angular UI libraries include Angular Material, PrimeNG, and Ng Zorro. However, not all your application code can be developed using a UI library, and sometimes your employer develops its own UI components; in these scenarios, you need to apply WCAG yourself. The first thing you need to do to ensure you adhere to WCAG is to use semantic HTML.

Semantic HTML is when you use the appropriate HTML element for the element you want to visualize on the screen. Often, developers like to use too many `<div>` and `<span>` elements; these elements don't tell a screen reader anything about their purpose. Try to use HTML elements such as `<label>`, `<button>`, `<input>`, `<header>`, `<footer>`, `<section>`, `<article>`, `<form>`, etc. When using semantic HTML elements, screen readers can better explain the page to users who can't see the page. Besides implementing semantic HTML, you should ensure that the user can navigate the page using the keyboard.

Using the tabindex attribute

You can ensure HTML elements are focusable with the tab key on your keyboard by adding a **tabindex** attribute on the HTML element. Certain HTML elements are inherently focusable, meaning they can receive keyboard focus without explicitly setting a tabindex attribute. This default focus behavior is implemented for the following elements: <a> or <area> with an href attribute, <button>, <iframe>, <input>, <object>, <select>, <textarea>, <SVG>, and <summary> elements that provide a summary for a <details> element. Developers don't need to manually add a tabindex attribute to these elements unless they want to alter their default focus behavior.

For instance, setting a negative tabindex value would remove the element from the focus navigation order, effectively making it non-focusable. However, it's essential to exercise caution when modifying the default behavior of focusable elements to ensure an intuitive and accessible user experience.

With the tabindex attribute, you can make HTML elements keyboard-focusable, prevent elements from being keyboard-focusable, and determine the focus order. As mentioned before, when you provide a negative integer to the tabindex attribute, the HTML element becomes non-focusable when using the keyboard. If you provide a 0 for the tabindex attribute, the element maintains the default tab order (from top to bottom, based on the order of the HTML elements). Lastly, you can provide positive integers. Elements with a positive integer will be focused before the default focus order kicks in, starting at tabindex 1 and ending at the highest tabindex. When there are no more tabindex attributes with a positive integer, the next focused element is the first HTML element with the default tabindex attribute:

```
<button type="button">1st default focus</button>
<button type="button">2nd default focus</button>
<button type="button" tabindex="100">Second focused</button>
<div tabindex="0">3rd default focus</div>
<button type="button" tabindex="1">First focused</button>
```

In the preceding code example, we provided three elements with a tabindex. The first element that will be focused is the button on the bottom with tabindex 1; after that, the button with tabindex 100 will be focused, and after that, elements will be focused from top to bottom (skipping the two elements we already focused because they have a positive tabindex attribute).

Now that you know how to control the keyboard focus of HTML elements and when to use the tabindex attribute, let's learn about ARIA attributes.

Adding ARIA attributes

Accessible Rich Internet Applications (ARIA) attributes are used to provide or add semantic meaning to HTML elements. In Angular, you can provide static or dynamic values for your ARIA attributes. Dynamic attributes might be useful if you want your ARIA attributes to be translatable or when they change for specific HTML elements based on some user interaction:

```
<button [attr.aria-label]="dynamicValue">…</button>
<button [attr.aria-label]="'translationkey' | translate">…</button>
<button aria-label="static value">…</button>
```

In the preceding code, you see the three options:

- First, we provide a dynamic value to the aria-label. In the first example, dynamicValue is a component property that can have different values over time.

- Then, you see how you can use a translated value for the aria-label.

- Lastly, we provided a static value for the aria-label.

There are many different ARIA attributes; let us take a look at the few most commonly used ones.

The aria-label

The aria-label attribute is by far the most used ARAI attribute. The aria-label attribute can be used to provide a text explanation for visual elements that don't have any text or to provide a more descriptive explanation of an element that does include text. For example, if you have a button HTML element that is styled as a hamburger menu, the button will not include any text. Without the aria-label attribute, a screen reader can't make anything of this, so you need to provide the aria-label attribute with an explanation:

```
<button aria-label="Hamburger menu" class="menu"></button>
```

In the preceding example, you can see we added the aria-label attribute and provided it with the text Hamburger menu, so a screen reader can clearly explain what the element is about. Alternatively, you can use the aria-label attribute to provide a better explanation for elements that do contain text:

```
<button aria-label="Submit add expenses form">Submit</button>
```

In the preceding example, we have a button containing the Submit text. By default, a screen reader will read the text of the HTML element, but only reading out Submit might not give the user enough information about what they are submitting, especially if you can't see the screen. Adding an aria-label attribute can provide a better explanation for the screen reader to read aloud to the user.

The role attribute

Another important and commonly used ARIA attribute is the `role` attribute. The `role` attribute in HTML defines an element's specific role or purpose within the web page or application. It can be added to various HTML elements to convey their intended purpose to assistive technologies, such as screen readers, which rely on this information to provide a meaningful experience to users with disabilities.

The `role` attribute can be added for most HTML elements, including but not limited to the following:

- Semantic elements such as `<nav>`, `<article>`, `<section>`, `<header>`, `<footer>`, and `<main>`.
- Form elements such as `<input>`, `<button>`, `<select>`, and `<textarea>`.
- Interactive elements such as `<a>` (anchor links), `<button>`, and `<option>` (within `<select>`).

The following is an example of the `role` attribute:

```
<a href="#" role="button">...</a>
```

In the preceding example, you can see we added a role attribute to an `<a>` tag, designating the role of the HTML element as a button. Commonly, `<a>` tags are used as links, but in this case, we indicate that it is used as a button instead.

> **More information**
>
> Besides the `role` and `aria-label` attributes, there are many other ARIA attributes, for example `aria-hidden`, `aria-checked`, `aria-disabled`, and `aria-readonly`.
>
> You can find a full list of all ARIA attributes and a detailed explanation at the following URL:
>
> `https://developer.mozilla.org/en-US/docs/Web/Accessibility/ARIA/Attributes`.

Now you know what the different accessibility guidelines are and how to implement them into your Angular applications. You learned why using a semantic HTML structure for your pages is important, so tools such as screen readers can better navigate and understand your components and pages. You learned how you can make elements focusable with the tab key and how you can control the focus order of different elements on the page. Lastly, you learned about the ARIA attributes and how they can be used to provide additional information for assistive technologies, such as screen readers.

Summary

In this chapter, you learned how to make your applications more accessible to people who speak different languages or are located in various locations. You learned about the Transloco library and how it can be used to implement localization and internationalization. You created language files to provide translation key-value pairs and implemented the translations in your HTML templates. You learned to translate values using structural directives, attribute directives, pipes, and the `TranslocoService`. After learning about translatable content, you learned how to format values for users from different locations. You learned about the `translocoCurrency`, `translocoDate`, and `translocoDecimal` pipe. You've seen how to configure your applications for localization and how to overwrite your default settings for specific instances within your code. After internationalizing and localizing your website, you learned about accessibility. You got familiar with the WCAG and how you can ensure they are implemented within your Angular applications.

In the next chapter, you will learn how to write unit and end-to-endtests for your Angular applications.

11
Testing Angular Applications

Writing automated tests for your applications is just as important as writing the application code. Many developers don't like to write tests or skip them altogether because they feel it's too time-consuming, but as your applications and workspace grow, having automated tests becomes ever more critical. When working on an extensive application, the chances are significant that your changes will impact many things throughout the application. Small changes can affect many things, which becomes even more apparent when you're making changes in a library used in many applications. You'll often find yourself in a scenario where you make changes and don't even know every application surface your changes will impact. Because you don't want to break features, you don't want to work on or manually test the entire workspace each time you make a change; you need automated tests that can test all affected code for you. Automated tests will help you look at your code differently; they can help you write better, more sturdy code. Automated tests will also catch bugs at an early stage and should give you the confidence to safely release code changes to production once all tests have successfully passed.

This chapter will dive into different types of automated tests and their purpose within your Angular applications. Next, you will dive deeper into the topic of unit testing and get some hands-on experience by writing unit tests for our Nx monorepo using Jest. Lastly, you will learn more about end-to-end testing and gain some experience writing end-to-end tests using Cypress. By the end of this chapter, you will understand why you need automated tests and how to write them for your Angular applications.

This chapter will cover the following topics:

- Different types of application testing
- Unit testing of Angular applications using Jest
- End-to-end testing of Angular applications using Cypress

Different types of application testing

In a world where software plays an increasingly important role and the applications we build are growing more complex, automated testing is becoming more critical. Companies are constantly looking to improve their applications to give users a better experience. To achieve this, many companies aim for continuous delivery of their software, meaning their updates can be automatically released to production at any given time. To ensure you can safely release updates without breaking things in production, you need automated tests that can run in your build pipelines to automatically test your software before releasing changes to your testing, acceptance, and production environments.

As applications become more complex, manually testing all changes becomes too time-consuming, and the chances of not testing something affected by your code changes increase significantly. Manual testing is also much slower, repetitive, and boring. Tasks that are time-consuming, tedious, and repetitive tend to get skipped and lead to mistakes. All that manual labor is also a considerable expense for the business, so having a sound automated testing system in place is a necessity.

Besides speeding up the testing process and making it less error-prone, automated tests should give you the confidence that any code you merge will not break the existing application code. If you've ever merged a large code change in an environment without a good testing suite, you know what a nerve-wracking experience it is and that you never really feel confident that your changes didn't break anything. If there are tests, they will catch bugs and help you think differently about how you implement your code, but having the ability to release changes confidently is the real goal we're trying to achieve when writing automated tests.

When it comes to automated tests for your Angular applications, you can divide them into four major types:

- Unit tests
- End-to-end tests
- Component tests
- Integration tests

Let's learn about these four types of tests, how they are used in Angular applications, and the differences between them.

Understanding unit tests

One of the fundamental aspects of software testing is **unit testing**. Simply put, unit tests validate small units of code, commonly individual functions, properties, or methods. Unit tests are used to test the update behavior of properties and the implementation of functions under different scenarios. Given a specified input, you expect the function to return a specific value and update certain properties. Unit tests run in isolation from the rest of your application, so you can test small units of code without them being affected by other parts of the application code; that way, you can quickly identify whether the function works as intended based on its implementation.

A common and popular technique for developing applications is **test-driven development** (TDD). In simple terms, when you use TDD to develop your applications, you first write the test scenarios and then the code implementations. Developing your code this way allows you to look at your code implementations from another perspective. Writing tests helps you to look at your code implementations differently in general. Still, when you first write all the possible test scenarios you want to cover and write the code implementation afterward, it changes your perspective.

In Angular applications, unit tests are commonly implemented using frameworks such as **Jest** or **Karma** and typically test specific Angular components, services, pipes, or directives. These tests are essential for verifying that each unit of code behaves as expected, adhering to its defined specifications and requirements. By isolating each unit of code, developers can identify and address bugs and issues early in development, promoting a more robust and stable application.

The primary purpose of unit tests in Angular applications is to give developers confidence in their code implementations, ensuring properties are updated and functions work as expected. By thoroughly testing individual units of code under different scenarios, developers can ensure that each line of code functions as intended, even as the codebase evolves and changes over time. This confidence is crucial to enabling developers to make changes and enhancements to the application with the assurance that existing functionalities remain intact.

A characteristic of unit tests is that they are quick to run, allowing developers to run them multiple times during the development process, making it easy to identify unintended side effects and bugs early. Unit tests also aim to cover a specific percentage of the codebase. Typically, companies like to test between 80% and 100% of the lines of code, functions, and branches (or paths) of the code; most unit test frameworks can enforce these thresholds, so you can't merge the code if you don't have enough testing coverage.

To summarize, unit tests are used to test small units of code such as functions, methods, and properties. With unit tests, you test code implementations under different scenarios to give you confidence that the code behaves as expected given a specified input. Commonly, you try to achieve a code testing coverage of between 80% and 100% for unit tests; so, compared to the other testing types (end-to-end, component, and integration), your unit tests will have the most test cases. Unit tests are fast to run and rarely fail because of your environment because they run in isolation.

Now that you have a good idea of what unit tests are and why they are useful, let's dive into the next type of tests: end-to-end tests.

Understanding end-to-end tests

End-to-end (e2e) tests are an integral part of the testing strategy for Angular applications. They offer a comprehensive approach to validating the application's behavior and functionality from the user's perspective. Unlike unit tests, which focus on testing individual units of code in isolation, e2e tests simulate real user interactions with the application, spanning multiple components and services to ensure that the application functions correctly as a whole.

In the context of Angular applications, e2e tests are commonly implemented using frameworks such as **Cypress**, **Playwright**, or **Protractor**. These frameworks provide helpful tools to automate browser interactions, allowing them to simulate user actions such as clicking buttons, entering text, and navigating between pages. By automating these interactions, developers can thoroughly test the application's user interface and workflow, identifying and addressing issues that may arise during real-world usage.

The primary goal of e2e tests is verifying that the applications work as expected from the perspective of the user, encompassing the application's functional and non-functional aspects. e2e tests serve and render the application (or specific libraries or modules of the application) in a real browser (you can also run them headless without opening a browser), visit a specific URL, and interact with the application as a user would. With e2e tests, you are testing whether components render correctly and whether features such as form submission, data retrieval and display, models, and error handling work as intended. By testing the application end-to-end, developers can ensure that multiple components and services work together seamlessly to deliver a cohesive user experience.

One of the main advantages of e2e tests is their ability to detect issues that may be absent when testing individual units of code in isolation. By exercising the entire application stack during the tests, including the frontend user interface, external dependencies, and (optionally) backend services, e2e tests can uncover issues related to data flow, communication between components, and interoperability with third-party services. This holistic approach to testing helps developers identify and address potential bottlenecks and failure points within the application, leading to a more robust and reliable software product.

However, while e2e tests offer many benefits, they also come with certain challenges and considerations that developers must address. The setup for e2e tests is more challenging than unit tests, and e2e tests are more prone to fail because of issues in the test environment. Nx already handles most of the setup for us, making starting with our e2e tests easier.

With e2e tests, you also don't have an easy way to detect code coverage, so they require more planning and coordination to ensure you're testing everything within your application and handling different scenarios and use cases.

Furthermore, e2e tests can be more time-consuming and resource-intensive to execute than unit tests due to their reliance on browser automation and the need to simulate real user behavior. As a result, developers must strike a balance between the depth and scope of e2e test coverage and the practical constraints of test execution time and resources.

Despite these challenges, e2e tests play a crucial role in ensuring Angular applications' overall quality and reliability, complementing other testing techniques such as unit tests and integration tests. By thoroughly testing the application from end to end, developers can gain confidence in its behavior and functionality, identify and address issues early in the development process, and ultimately deliver a high-quality user experience to their customers.

To summarize, e2e tests are designed to test your application from the user's perspective and interact with your application in a real browser. They ensure your application (or specific libraries or modules) works as a whole and responds in the intended way to user interactions. e2e tests, while more time-consuming to write and execute than unit tests, provide the assurance that the user can interact with your applications as you intended, fostering a stronger connection with your end users.

Now that you know what e2e tests are and how they differ from unit tests, I will briefly explain component and integration testing.

Understanding component tests

Component testing is a relatively new concept compared to unit and e2e testing. In modern frontend frameworks such as Angular, we develop applications using **components**. Components can be simple components, such as buttons, or more complex ones, such as tables or forms.

Using component tests, frameworks such as Cypress provide a new approach for testing component-based applications. Instead of visiting a URL and running the entire application, component tests mount individual components and test those components in isolation. Component testing is like unit testing of e2e testing. You still mount the component and show it in a browser to interact with the component like a user would, but you test it in isolation from the rest of your application. Testing components in isolation allows you to test the component from a user interaction perspective without worrying about the rest of the application.

One thing to keep in mind is that even if all your component tests are passing, it does not automatically mean your application is working as expected. Components can work in isolation but fail when they're combined or have to interact with other components in your application. Compared to e2e tests, component tests don't need the entire system to be executed, so they can run faster and rarely fail due to issues with your test environment.

If you and your team want to implement component tests, it is up to you; these tests can help reduce the number of e2e tests you need to write. I like to write more e2e tests as opposed to component tests. Component tests still need to be adopted as an industry standard; most companies only require unit, e2e, and integration tests.

To summarize, component tests test individual components from the user's perspective. Component tests ensure a component works in isolation but don't ensure the component works within the context of your entire application. Component tests are easier to set up than e2e tests as they don't need to run the entire application. You can think of component testing as a mix between unit and e2e testing.

Now that you know what component tests are and how they differ from e2e tests, we will finish this section on different testing types by explaining integration tests.

Understanding integration tests

Integration tests are used to test whether different modules and elements of your software integrate without breaking. They are generally the final testing stage before you release your changes to production. So, unit tests focus on testing individual units of code in isolation, end-to-end tests simulate and test user interactions for specific application libraries and modules, and integration tests are used to test the interactions between various modules and elements within the application.

Integration tests can be used and written for different integration levels within an application. For example, you can write functional tests comparable to unit tests. Still, instead of testing code implementations for an isolated component or service, you test whether your functionality and implementations work as expected for a group of components or services that work together. You can also write integration tests from the user's perspective, similar to e2e tests. When you write integration tests from the user's perspective, you can test whether your frontend works together with your API or whether your deployed application is composed of multiple Angular applications. You can also test whether the different applications can work together when everything is deployed.

When you create integration tests from the user's perspective, you commonly run the tests in an environment that mirrors the production environment. You test with an actual deployed application with real APIs and data. By testing on a deployed system, you can test whether all elements of your software work together as expected without any boundaries.

Now you know about unit, e2e, component, and integration tests, it's time to get our hands dirty and write some tests ourselves. We will skip the integration tests because we don't have a large system or a deployed version with different elements that integrate. We will start by writing and running unit tests for our Angular application using the Jest testing framework. After writing our unit test, we will finish the chapter by writing e2e tests using the Cypress testing framework.

Unit testing of Angular applications using Jest

When you add an Angular project to your Nx monorepo, the application is set up to use Jest as a test runner by default. Jest is a testing framework commonly used to write and run automated unit tests for JavaScript and TypeScript-based applications. This section will give you hands-on experience writing unit tests for your Angular application using Jest. Before you start writing tests, let's expand upon the default configuration Nx provided to make your testing experience better.

Setting the coverage threshold

The first thing you want to add in the Jest configuration is a coverage threshold for the minimum required percentage of lines, functions, and branches that unit tests should cover. A commonly used percentage is 80%, but you can set the coverage percentages to whatever you and your team deem enough to make you confident that new changes won't break existing code. You can add the global configurations for testing coverage inside the `jest.preset.js` file in the root of your Nx monorepo. Additionally, you can set specific configurations for each project in the `jest.config.ts` file at the root of each project. I will only add the following configurations in `jest.preset.js` in the root of the Nx monorepo:

```
coverageThreshold: {
  global: {
      lines: 80,
      functions: 80,
      branches: 80
  },
},
collectCoverage : true,
coverageReporters: [
  "cobertura",
  "lcov",
  "text",
]
```

The preceding configuration ensures that all branches, functions, and lines have a minimum test coverage of 80%. The configuration also tells Jest to collect the coverage results and present you with a text-based coverage report. In the coverage report, you can see how much of your code is covered, what lines, functions, and branches are missing, and their respective page line numbers.

Now that you have configured your test coverage reports, it's time to add a testing module for Transloco.

Adding additional configurations

The testing module makes it easy to import the correct configuration to test components using Transloco. In the root of your `expenses-registration` project, you can create a `transloco-testing.module.ts` file and add the following content:

```
import { TranslocoTestingModule, TranslocoTestingOptions } from '@
ngneat/transloco';
import en from '../assets/i18n/en.json';
import nl from '../assets/i18n/nl.json';


export function getTranslocoModule(options: TranslocoTestingOptions =
{}) {
```

```
    return TranslocoTestingModule.forRoot({
      langs: { en, nl },
      translocoConfig: {
        availableLangs: ['en', 'nl'],
        defaultLang: 'en',
      },
      preloadLangs: true,
      ...options
    });
}
```

In the preceding example, we created a `getTranslocoModule()` function. This function will be used inside our unit test files to add the necessary Transloco configuration for the test setup. It's simply a function returning the `TranslocoTestingModule` class provided by the `Transloco` library. At the top of the file, we import two JSON files containing our translations. If you want to import these two JSON files without trouble, you need to add the following configurations inside your `tsconfig.base.json` file:

```
  "resolveJsonModule": true,
  "esModuleInterop": true,
```

After adding the preceding configuration and the `transloco-testing.module.ts` file with the `getTranslocoModule()` function, we are almost done with our addition to the default Jest setup Nx provided us with. Lastly, we need to update the `transformIgnorePatterns` configuration inside our `jest.config.ts` files to the following:

```
  transformIgnorePatterns: ['node_modules/?!(.*\\.mjs$|@ngneat)'],
```

When you change the `transformIgnorePatterns` configuration, you ensure Jest will not start to complain about missing imports and packages inside your `node_modules` folder. There already are `transformIgnorePatterns` configurations inside each `jest.config.ts` file in your Nx monorepo, but in many cases, you need to adjust them or else your tests might fail based on things inside your `node_modules` folder.

That is all the additional setup we will be doing. You can always add additional configurations as needed. You can find all additional Jest configurations in their official documentation at `https://jestjs.io/docs/configuration`.

Now that you have added the configuration needed to test Transloco and obtained testing coverage reports, let's start to write and run unit tests for our Angular application.

Writing and running unit tests

You write your unit tests inside **spec** files using the Jest testing framework. When we created our Nx monorepo, Nx already configured everything we needed to write and run tests using Jest. Additionally, every time we used the Nx generator to create a component, service, pipe, or directive, Nx created a `.spec.ts` file for the created resource. These `.spec.ts` files contain the default-generated unit tests.

Let's take the `expenses-registration` Angular application as our example. When you generated the application, Nx generated an `AppComponent` class for you and an `app.component.spec.ts` file where the default generated unit test for `AppComponent` resides. Additionally, we created `ExpensesOverviewPageComponent` and `ExpensesApprovalPageComponent` inside the *expenses-registration application*; for both these components, Nx also generated `.spec.ts` files. Let's start with these files.

Fixing the generated spec files

If you currently run the tests inside these spec files, they will fail. The tests will fail because we haven't touched the spec files since they have been generated, but we did adjust the component classes. So before we try to run the tests, let's fix the spec files one by one. We will also write some new tests and explain what Nx generated for us. Starting with the `app.component.spec.ts` file, let's see what Nx has generated inside the spec file:

```
describe('AppComponent', () => {
  beforeEach(async () => { ...... });
  it(<should render title>, () => { ...... });
  it(`should have as title <finance-expenses-registration>`, () => {
...... });
});
```

Nx generated the preceding code for you. As you can see, there are `describe()`, `beforeEach()`, and two `it()` functions. All of these functions have some additional code inside their respective callback functions, but we will ignore that for now. Let's first explain what `describe()`, `beforeEach()`, and `it()` functions are used for:

- `describe()`: The `describe()` function is used to group multiple tests together and describe what element we are testing. You provide the `describe()` function with two parameters: a string with a description of what we're writing tests for—in our example, `AppComponent`—and a callback function where we will write out specific test cases.

- `beforeEach()`: The `beforeEach()` function is used to perform specific steps before each test, commonly configurations such as setting up `TestBed`, creating the component, service, pipe, or directive we are testing, and any additional configuration we want to do before each test we run.

- `it()`: The `it()` functions define each test case. An `it()` function takes in two parameters: a string containing a description of the test case and a callback function containing the testing logic.

Now that you know what `describe()`, `beforeEach()`, and `it()` functions are, let's replace the generated code of your `app.component.spec.ts` file with something that reflects the current state of the app component.

Defining our test cases

Before we start adjusting the code inside the spec file, let's first clarify what we want to test.

If you look inside your `AppComponent` class, you find two properties: a `translationService` and a `navItems` property. Additionally, inside the HTML template of your app component, you'll find the navbar component with some inputs and an output for when the selected language changes. As we mentioned before, when writing unit tests, you want to test a single unit of code in isolation—in this case, our app component. So, what functionalities are related to the app component?

- Defining the component class properties.

- Calling the `setActiveLanguage` method on the `translationService` property of the navbar to emit a `languageChange` event.

The aforementioned points are the only component logic related to the app component; checking whether the navbar inputs are correctly handled and whether the navbar renders correctly is logic related to the navbar and should be tested in the spec file of the navbar component. If we were to check these things inside the spec file of the app component, we would be testing whether the navbar and app components integrate correctly. For the same reason, we do not check whether `TranslationService` actually adjusts the active language after we call the method. This would test the integration between the app component and `TranslationService`. From the perspective of the app component, we are only interested if the app component actually makes the function call. Now we know what we will be testing, let's define what `it()` statements we will be creating inside our spec file:

- They should create the component and set the component properties with the expected values.

- They should call the `setActiveLanguage` method when the `languageChange` event is emitted.

Now that we have defined the `it()` statements we're about to define inside the `app.component.spec.ts` file, let's start to adjust the file step by step so we can actually test these statements successfully.

Adjusting the code inside the spec file

We defined the test cases we wanted to write and learned about the three main functions inside the spec files. Now, let's write our test cases and learn how to configure your testing modules, make assertions for your test cases, and actually run the tests.

The describe() function

We will start by removing all the code inside the `describe()` function so we can start fresh. The `describe()` function itself can remain as it was generated for you. After you remove the generated code, start by defining three properties inside the `describe()` function:

```
let component: AppComponent;
let fixture: ComponentFixture<AppComponent>;
const mockTranslationService = {
  setActiveLanguage: jest.fn(),
  getLanguages: jest.fn().mockReturnValue([]),
};
```

As you can see in the preceding code snippet, we added `component`, `fixture`, and `mockTranslationService` properties inside the `describe()` function. The `component` variable will hold an instance of our `AppComponent` class, `fixture` will be an element containing a test harness that can be used to debug and interact with the app component (the class, native element, element ref, lifecycle methods, etc.), and `mockTranslationService` will be used as a value for the `TranslationService` injectable we use inside the app component. We use this mock version of `TranslationService` to simplify the setup we need to do inside our spec file. Because we don't want to test the integration between our app component and `TranslationService`, we want to test the app component in isolation. After you have defined these three properties, it's time to add the `beforeEach()` function.

The beforeEach() function

The `beforeEach()` function will be added underneath the three properties we added just now and will be used to configure `TestBed` and assign our `component` and `fixture` properties before each test. Let's start simply by defining the `beforeEach()` method itself:

```
beforeEach(async () => {});
```

Now inside the callback of the `beforeEach()` function, we start by configuring the testing module using the `TestBed.configureTestingModule()` method. The testing module requires everything needed to create our app component:

```
await TestBed.configureTestingModule({
  imports: [AppComponent, RouterTestingModule, getTranslocoModule()],
  providers: [{
      provide: TranslationService,
      useValue: mockTranslationService,
  }]
}).compileComponents();
```

As you can see in the preceding code, we need to import three classes and define a provider to configure the testing module. You need to import the `AppComponent` class because `AppComponent` is a standalone component, `RouterTestingModule` because we use the `RouterOutlet` inside the app component template, and `TranslocoTestingModule` using the `getTranslocoModule()` function we defined in the *Adding additional configurations* section of this chapter. Besides the imports, you need to create a provider for `TranslationService` so that the app component uses `mockTranslationService` during the tests. At the end of the `configureTestingModule()` method, you need to call the `compileComponents()` methods so that Jest will compile everything we defined inside the testing module configuration.

After the `TestBed` testing module is configured, we will assign the `component` and `fixture` properties. The `fixture` property will be assigned using the `TestBed.createComponent()` method. Calling the `createComponent()` function on `TestBed` will freeze the current `TestBed` class, meaning you can't call `TestBed` configuration methods anymore. It will also return a test harness that can be used to interact with the component created inside your test cases:

```
fixture = TestBed.createComponent(AppComponent);
```

After assigning `fixture`, you assign the `component` variable using the `componentInstance` property of `fixture`. This `componentInstance` property is an object containing all the properties and functions of the component you're testing—in our case, `AppComponent`:

```
component = fixture.componentInstance;
```

Lastly, you need to call the `detectChanges()` method on `fixture` so that change detection will run for the created app component:

```
fixture.detectChanges();
```

Now that you have defined the `beforeEach()` function and configured `TestBed`, we can start with our first `it()` function and define the first test case.

The first it() function and test case

You can define your `it()` function underneath the `beforeEach()` function. In our case, the first test case should test whether the component is successfully created and the `navItems` and `translationService` properties are correctly assigned:

```
it('should create the component and set the component properties with
the expected values', () => {
  expect(component).toBeDefined();
  expect(component.navItems).toEqual([{ label: 'expenses approval',
route: '/expenses-approval' }]);
  expect(component[<translationService>]).
toEqual(mockTranslationService);
});
```

As you can see in the preceding code, we start with the `it()` function and provide the function with a description. Then, in the callback function, we assess what we want to test using the `expect()` function combined with an assertion method. You provide the `expect()` method with the value you want to test and expect to be or not to be something. In our case, we first expect the component property (which is assigned without a component instance inside the `beforeEach()` function) to be defined. Next, we expect the `navItems` property of the component to equal the object we defined inside the component class for the `navItems` property. Lastly, we expect `translationService` to equal `mockTranslationService`.

Now you have defined the `beforeEach()` function, configured `TestBed`, created the component inside the `beforeEach()` function, and written your first testing case. You can run the test inside the `app.component.spec.ts` file. You run unit tests by running the following command in the root of your Nx monorepo:

```
npx nx run <project-name>:test
```

In the preceding Terminal command, you need to replace the `<project-name>` placeholder with the actual name of the project you want to run tests for. You can find the project name inside the `project.json` file of each application or library within your Nx monorepo. To run the unit tests for a specific file, add the `-test-file` flag at the end of the command. For example, to run the unit tests for our `app.component.spec.ts` file, you run the following command:

```
npx nx run finance-expenses-registration:test --test-file=app.
component.spec.ts
```

The preceding command will run the tests inside the `app.component.spec.ts` file. After running the tests, you'll notice that your test case is failing with the following error message: **Can't bind to 'navbarItems' since it isn't a known property of 'bt-libs-navbar'**. This is because we're using Angular signal inputs inside `NavbarComponent` and Jest doesn't support signal inputs (at the time of writing).

As a workaround, you can create a simplified replica of `NavbarComponent` that is used for unit testing components that use `NavbarComponent` in their template. Such a replica is commonly named a **stub**. Using a stub component will not only fix the signal input issue but also ensure that you test the `AppComponent` functionality in isolation instead of having it integrated with `NavbarComponent`. You create the navbar stub component in your `common-components` library inside the `navbar` folder by adding a `navbar.component.stub.ts` file with the following content:

```
@Component({
  selector: "bt-libs-navbar",
  standalone: true,
  template: <>,
})
```

```
export class StubNavbarComponent {
  @Input() navbarItems = [];
  @Input() languages = [];
  @Output() languageChange = new EventEmitter();
}
```

After creating the stub component, export it in the `index.ts` file of your `common-components` library so you can access the stub component in your spec files. Now, inside the `beforeEach()` function of the `app.component.spec.ts` file, you can ensure the app component uses the stub navbar component during the tests instead of the regular navbar component. You can achieve this by changing the navbar component import for the stub navbar component using the `TestBed.overrideComponent()` method. You need to simply remove the `NavbarComponent` import and add the `StubNavbarComponent` import:

```
TestBed.overrideComponent(AppComponent, {
  add: {
    imports: [StubNavbarComponent],
  },
  remove: {
    imports: [NavbarComponent],
  },
});
```

As you can see in the preceding code, we remove the `NavbarComponent` import from `AppComponent` and add `StubNavbarComponent`. It's important that you overwrite the component imports before you call the `TestBed.createComponent()` method and freeze `TestBed`; otherwise, your override will not be included in `TestBed`.

Using stub components and services can be helpful in cases like this, where Jest still needs to add support for specific features. Additionally, stubs ensure that you're not integration testing but focusing on units of code in isolation. If you want to unit test the navbar, for example, you should do so in the spec file of the navbar component and not in the spec file of the app component. Additionally, using stub components can simplify the setup you need in the `beforeEach()` method to ensure Jest can create the component or service you want to unit test.

Suppose you rerun the app component unit tests after adding the stub navbar component. In that case, you'll see that the test case we defined in the spec file is passing, meaning the test successfully creates `componentInstance`. Still, the test run fails because we don't meet the configured coverage requirements of 80%. If you look at the coverage report in your terminal, you'll see that the coverages of your `app.component.ts` and `app.component.html` files are 100%, but the coverage of your `translation.service.ts` file is 0%, bringing the total testing coverage under the required 80%.

So, why is the testing coverage including `translation.service.ts`, and should you care? The `translation.service.ts` file is included in your coverage report because, by default, Jest (and other test runners) will consist of all files in the coverage report you import and use within the class you're testing—in this case, the `AppComponent` class. Should you care, and do you need to fix this shortfall in your coverage percentage?

The answer is it depends on how you run the tests. If you're running the tests for a single file, as we are doing now, you shouldn't care and should only focus on the files related to the unit you're testing—in our case, the app component. After all, you want to write shallow unit tests that only test a single unit of code in isolation, so if you have a testing coverage of 80% or higher for the files related to the unit you're testing, everything is good! However, if you're running the unit test for an entire project by omitting the `-test-file` flag, you should care about the coverage percentage. For your entire project, you should have enough coverage. In this example, the code related to the `translation.service.ts` file should be tested in a `translation.service.spec.ts` file. If you run the unit tests for the entire project and you cover the logic of your `translation.service.ts` file inside the `translation.service.spec.ts` file, you will not have the shortfall in your coverage report and the test run will succeed. Now that we have clarified that, let's add our second `it()` function for `app.component.spec.ts`.

The second it() function

While we have 100% testing coverage for our app-component-related files, our test might not give us the confidence we need that everything works as expected. We don't test whether the app component class calls the `setActiveLanguage` method when the navbar emits a `languageChange` event, so let's add a test for this. You can add the following code to test whether the app component calls the `setActiveLanguage` method when it receives the `languageChange` event:

```
it('should call the setActiveLanguage method when the languageChange
event is emitted', () => {
  const setActiveLanguage = jest.
spyOn(component['translationService'], 'setActiveLanguage');
  const navbarElement = fixture.debugElement.query(By.
directive(StubNavbarComponent));
  navbarElement.triggerEventHandler('languageChange', 'nl');
  expect(setActiveLanguage).toHaveBeenCalledWith('nl');
});
```

In the preceding code, quite a lot is happening, so let's examine each line carefully.

First, we define the `it()` function and provide it with a description of our test case. Inside the callback function, we start by creating a spy element.

Spy elements are commonly used to spy on a specific function and check whether the spied-on function is called, how many times, and with what parameters. We will use our spy element to check whether the setActiveLanguage method of the translationService property is called. Create the spy object by using the jest.spyOn() function. Inside the jest.spyOn() function, first provide the object containing the function you want to spy on and then, as a string, the function name you want to spy on.

After creating the spy object, we use debugElement of fixture to access StubNavbarComponent inside our HTML template and save it in a constant named navBarElement. Next, we use the triggerEventHandler method on navBarElement to trigger the languageChange event and provide nl as the event data.

After triggering the languageChange event, we expect that setActiveLanguage() is called with the nl parameter. We check whether this is correct by providing the expect() function with the setActiveLanguage spy object and calling the toHaveBeenCalledWith('nl') assertion method on the expect() function.

After adding your second testing case to the app.component.spec.ts file, you can run the tests again, and you'll notice that both test cases are successful.

To summarize, you learned that the describe() function is used to group test cases and the beforeEach() function is used to configure TestBed and define values before each test case runs. The it() functions are used to define your test cases, and inside the it() functions, you use the expect() function combined with assertion methods to perform your test statements. You can create and use spy objects to validate whether functions are called. When writing unit tests, you should write shallow unit tests that focus on a single unit of code instead of testing the integration of different components and services. You can write shallow unit tests and prevent issues with unsupported features by creating stub components and services, which are simplified replicas used within your unit tests.

Now that you have a better understanding of unit tests and you've created your first tests, we will fix the additional spec files of the expenses-registration component so we can do a successful test run for the application. Additionally, you will learn how to run unit tests for multiple projects in your Nx monorepo.

Adding additional unit tests for the expenses-registration application

We will now start writing some additional unit tests for the *expenses-registration application* so you can successfully run the unit tests for the entire application without specifying the -test-file flag. To start, we will remove the expenses-approval-page.component.spec.ts file because we haven't added any code yet inside the expenses approval component. After removing expenses-approval-page.component.spec.ts, we will adjust the tests inside expenses-overview-page.component.spec.ts. We made quite some adjustments to the ExpensesOverviewPageComponent class, so fixing the related spec file will be a bit more work compared to the spec file for AppComponent.

Let's start by running the unit tests with the following command and see what pops up:

```
npx nx run finance-expenses-registration:test --test-file=expenses-
overview-page.component.spec.ts
```

It might come as no surprise that the test run fails. Let's fix the issues with the spec file one by one, starting with the import of `ExpensesOverviewPageComponent` inside the spec file.

Adjusting the code inside our other spec file

Because we changed the export of `ExpensesOverviewPageComponent` to a default export, we also need to adjust the import inside the spec file:

```
import ExpensesOverviewPageComponent from './expenses-overview-page.
component';
```

After changing the `import` statement, you need to adjust the spec file so `TestBed` can successfully create the `expenses-overview` component. Just like the app component, the `expenses-overview` component uses `TranslationService`, so we will make a mock object for this service (alternatively, you can create a stub service for it and also use that inside the spec file of the app component):

```
const mockTranslationService = {
  translocoService: { translate: jest.fn() },
  translationsLoaded: signal(false) as WritableSignal<boolean>,
};
```

As you can see in the preceding code, `mockTranslationService` for this spec file differs from the `mockTranslationService` class we created for the spec file of the app component. The mock objects differ because we only include what we use inside the component we are about to test in the mock object; in this case, the `expenses-overview` component only uses the `translocoService` and `translationsLoaded` properties of the service. Besides `mockTranslationService`, we also need a stub for `ExpensesFacade`. You can copy `StubExpensesFacade` from the GitHub repository for this book. The `expenses.facade.stub.ts` file is located next to the regular `expenses.facade.ts` file inside the finance `data-access` library. After creating the mock and stub objects we need for our unit tests, we can create the `beforeEach()` function and set up `TestBed`:

```
beforeEach(async () => {
  await TestBed.configureTestingModule({
    imports: [ExpensesOverviewPageComponent, getTranslocoModule()],
    providers: [
      { provide: ExpensesFacade, useClass: StubExpensesFacade },
      { provide: TranslationService, useValue: mockTranslationService,
},
      provideTranslocoLocale({
```

```
        langToLocaleMapping: { en: 'en-US', nl: 'nl-NL' }
      })
    ]
  }).compileComponents();

  fixture = TestBed.createComponent(ExpensesOverviewPageComponent);
  component = fixture.componentInstance;
  fixture.detectChanges();
});
```

As you can see in the preceding code, we import ExpensesOverviewPageComponent and TranslocoTestingModule using the getTranslocoModule() function. After the imports for the testing module, we added some providers that needed to configure the testing module. We provide ExpensesFacade and TranslationService with the mock and stub values, and we provide the TranslocoLocale configuration because we use the localization pipes inside the expenses-overview page. After configuring the imports and providers for the testing module, we created and assigned the fixture and component properties and called detectChanges() on fixture. Now we have configured everything needed to create the component inside TestBed, let's remove all it() functions and write our own test cases.

Writing the test cases

We will create the following test cases:

- The test should create the component and initialize the properties correctly.

- It should fetch expenses on init.

- It should translate the title if translations are loaded.

- It should change summaryBtnText if onSummaryChange is called.

- It should call addExpense on the expenses facade with the correct values when onAddExpense is called.

Now that we have defined our test cases, let's create them one by one:

```
it('should create the component and initialize the properties
correctly', () => {
  expect(component).toBeTruthy();
  expect(component[<expensesFacade>]).
toBeInstanceOf(StubExpensesFacade);
  expect(component[<translationService>]).
toEqual(mockTranslationService);
  expect(component.translationEventsEffect).toBeDefined();
  expect(component.expenses()).toEqual(component[<expensesFacade>].
expenses());
```

```
    expect(component.showAddExpenseModal()).toBeFalsy();
    expect(component.showSummary()).toBeFalsy();
    expect(component.summaryBtnText()).toEqual('Show summary');
});
```

As you can see in the preceding code, this test is really straightforward; we simply check whether the component variable is defined and whether each component property is initialized with the value we expect. As you may have noticed, we used some new assertion methods here, such as `toBeFalsy()` and `toBeInstanceOf()`. These assertion methods can be used to check whether a value is false in a Boolean context and whether an object is an instance of a specific class. Besides `toBeFalsy()` and `toBeInstanceOf()`, we did nothing new in this test, so let's move on to our next test case:

```
it('should fetch expenses on init', () => {
    const fetchExpenses = jest.spyOn(component['expensesFacade'],
'fetchExpenses');
    component.ngOnInit();
    expect(fetchExpenses).toHaveBeenCalled();
});
```

In the preceding test, we create a spy object to spy on the `fetchExpenses()` function of `ExpensesFacade`. After that, we call `ngOnInit()` for the component we are testing—in this case, the `expenses-overview` page component—and at the end of our test, we use the `expect()` function to check whether the `fetchExpenses()` function is called. As you can see, calling a method declared in the component we are testing is super straightforward; you simply use the `component` variable and call the method you want to run. Now that we've also covered this test, let's move on to the next test case we defined:

```
it('should translate title if translations are loaded', fakeAsync(()
=> {
    const translateSpy = jest.spyOn(component['translationService'].
translocoService, 'translate');
    expect(component[<translationService>].translationsLoaded()).
toBeFalsy();
    expect(translateSpy).not.toHaveBeenCalled();

    mockTranslationService.translationsLoaded.set(true);
    tick();
    expect(translateSpy).toHaveBeenCalledWith(<expenses_overview_page.
title>);
}));
```

In the preceding test, a bit more is going on, and we used some new techniques. Let's explore what we do here in more detail. Because we are testing a signal effect in this test and signal effect are asynchronous, we wrapped the callback of our `it()` function inside a `fakeAsync()` function. Inside the `fakeAsync()` function, times are synchronous. You can manually execute microtasks by calling `flushMicroTasks()` and simulating time passing with the `tick()` function. After using the `fakeAsync()` function, we first define a spy object. Then, we check whether the `translationsLoaded` signal has a false value and that the `translate()` function that we use inside our signal effect isn't called. Next, we set the value of the `translationsLoaded` signal to `true`. This should trigger the signal effect again, and this time, we should reach the part where we use the `translate()` function. Because the signal effect is asynchronous, we first call the `tick()` function to simulate the passing of time, and after that, we check whether our spy object is called with the correct translation key.

Now that we have explained the `fakeAsync()` and `tick()` functions used in our last test case, let's continue and add the next test case:

```
it('should change the summaryBtnText if onSummaryChange is called', ()
=> {
  expect(component.showSummary()).toBeFalsy();
  expect(component.summaryBtnText()).toEqual('Show summary');

  component.onSummaryChange();
  expect(component.showSummary()).toBeTruthy();
  expect(component.summaryBtnText()).toEqual('Hide summary');
});
```

As you can see in the preceding code, this is a simple test. We first check whether the `showSummary` signal is false and the `summaryBtnText` computed signal returns **Show summary**. Next, we call the `onSummaryChange()` function and check whether the `showSummary` signal and the `summaryBtnText` computed signal are adjusted correctly. After adding the preceding test, there is only one test case to add to our spec file:

```
it('should call addExpense on the expenses facade with the correct
values when onAddExpense is called', () => {
  const addExpense = jest.spyOn(component['expensesFacade'],
'addExpense');
  const expenseToAdd = { description: 'test', amount: { value: 50,
vatPercentage: 20 }, date: '2019-01-04', tags: ['printer'], id: 999 };
  component.onAddExpense(expenseToAdd);
  expect(addExpense).toHaveBeenCalledWith(expenseToAdd);
  expect(component.expenses().expenses).toContainEqual(expenseToAdd);
});
```

In the preceding test, we first create a spy object to spy on the `addExpense()` function of `ExpensesFacade`. After creating the spy object, we create an `expense` object to provide to the `onAddExpense()` method. After creating expense, we call the `onAddExpense()` method and provide it with the `expenseToAdd` property. After we call the `onAddExpense()` method, the `addExpense()` function of the facade should be called with the `expenseToAdd` property as a function parameter. We verify that the `addExpense()` function is called with the correct parameter using the `toHaveBeenCalledWith()` assertion method. Lastly, we use the `toContainEqual()` assertion method to check whether expense is added to the expenses signal of the `expenses-overview` page component.

After adding the last test case, you can run the tests again using the following command:

```
npx nx run finance-expenses-registration:test --test-file=expenses-
overview-page.component.spec.ts
```

After you run the tests again, you'll find that all tests are passing and you have 100% coverage for the files related to the `expenses-overview` component. You can write some additional tests to test the template, but this will also be covered with the e2e tests we will write in the next section. Now that you know how to run the unit tests for your individual spec file, let's examine how to run unit tests for one or more projects within your Nx monorepo.

Running unit tests for one or more projects

We ran the unit tests for the spec files one by one, so now let's run them for the entire `finance-expenses-registration` project. When you run the unit tests for an entire project, it will run the tests in all spec files found inside that Nx project. Note that this will not include any library projects you use inside the project. For example, to run the test for all spec files inside the `finance-expenses-registration` project, you use the following Terminal command:

```
npx nx run finance-expenses-registration:test
```

When you run the preceding command, you'll notice that the test run fails because we do not meet the coverage threshold of 80%. This is because we do not test the `translation.service.ts` file. As an exercise, you can create the spec file for `TranslationService` yourself; alternatively, you can lower the coverage threshold.

Besides running the unit tests for a single project, you can also run unit tests for multiple projects at the same time using the `run-many` command. When using the `run-many` command without any additional parameters, you will run the unit tests for projects found in your entire Nx monorepo:

```
nx run-many -t test
```

Furthermore, you can add specific project names at the end of the terminal command to only run the unit tests for specific projects:

```
nx run-many -t test -p proj1 proj2
```

You can also run the unit tests for all projects and exclude specific projects from the test run using the –exclude flag:

```
nx run-many -t test --exclude excluded-app
```

Lastly, you can use the affected terminal command. The affected command can be used to run the unit tests for all projects affected by your changes. Nx will look at its cache and check what projects have changed since the last time the unit tests ran. Any project affected by changes made after the latest cached test run will be run when using the affected command. The affected command is particularly useful when you run your unit tests on a build pipeline and want to test your code each time you're merging code:

```
nx affected -t test
```

To summarize, you can run tests for individual spec files, for one or more Nx projects, or for projects affected by your changes. Unit tests are meant to test isolated units and should make you confident that your changes will not break your existing code implementations. Unit tests consist of three main parts, the describe(), beforeEach() and it() functions, and inside your test cases, you assert them using the expect() function.

Now that you know how to write and run unit tests, it's time to dive deeper into the topic of e2e tests.

End-to-end testing of Angular applications using Cypress

When you create an application using the Nx CLI or Nx console, two projects are created for you: a regular application (in our case, an Angular application) and an e2e project configured to test the generated application project using the Cypress testing framework. For example, when we created the expenses-registration project, Nx also created an expenses-registration-e2e project. The folder for the expenses-registration-e2e project is located next to the expenses-registration project folder.

Before we start writing our own e2e tests, let's see what Nx generated for us inside the expenses-registration-e2e folder. When you open the expenses-registration-e2e folder, you find some folders and four files. The .eslintrc.json, cypress.config.ts, project.json, and tsconfig.json files are all meant to configure Cypress and the e2e project. We want to adjust one small thing inside the tsconfig.json file; you can leave the rest of the files untouched. Inside this tsconfig.json file, you'll find an include array; inside this include array, add the following string:

```
"src/**/*.cy.ts"
```

Besides the configurations created by Nx, you also want to add one small thing inside the `.eslintrc.json` file at the root of your Nx monorepo. Inside the `.eslintrc.json` file at the root of your Nx monorepo, you'll find a `project` array; inside this array, add the following value:

```
"apps/*/*/tsconfig.json"
```

Without the two aforementioned additions to your configuration, you will run into some ESLint parsing errors. After adding the additional configurations, let's see what else Nx has generated inside the `expenses-registration-e2e` project.

You'll see a `cypress` folder and an `src` folder inside the `expenses-registration-e2e` folder. The `cypress` folder can be ignored; inside the `src` folder, you'll find `e2e`, `fixtures`, and `support` folders that have the following purposes:

- `e2e`: Inside the e2e folder, you'll add the files containing your e2e tests. Nx already generated an `app.cy.ts` file inside this folder. As you can see, the file name ends with `cy.ts`. This is a naming convention for the files containing your Cypress e2e tests. The `cy` at the end of your file is short for Cypress.

- `fixtures`: Inside the `fixture` folder, you can add JSON files containing the mock data you want to use inside your e2e test. Using fixtures is useful when you want to use specific data for e2e tests. Additionally, you often won't have an API or mocking service you can or want to use during your e2e tests. Adding mocking services or APIs in your e2e tests often requires a lot of additional setup in both your local environment and in your pipelines where you want to run the e2e tests. Besides additional setup, using a real API or the same mocking service as your development environment can lead to more instability for your e2e test.

- `support`: Inside the `support` folder, you'll find everything you need to write and run your e2e tests. Some things you place inside the `support` folder are a file containing all the imports you use inside the e2e project, a file containing page objects used inside your `.cy.ts` files, additional setup files, or files with custom Cypress commands.

Now that you have an idea of what Nx generated for you and what the files and folders inside your e2e project are used for, let's start to write and run e2e tests for the `expenses-registration` project.

Writing your first e2e test

Start by removing the `app.cy.ts` file and replacing it with an `expenses-registration.cy.ts` file. Inside this `expenses-registration.cy.ts` file, we will write the e2e tests that will test the *expenses-registration application*. As we did with the unit tests, we define a `describe()` function. You use the `describe()` function for grouping multiple e2e tests, similar to the `describe()` function we used for the unit tests. The `describe()` function takes two arguments: a description and a callback function:

```
describe('finance-expenses-registration', () => {});
```

Now, inside the callback function of the `describe()` function, we will add a `beforeEach()` function. Inside the `beforeEach()` function, you can define the steps you want to perform before each e2e test. Some common steps defined inside the `beforeEach()` function are visiting the URL of your application, setting up interceptors, logging in as a user, and closing the cookie consent message. In our case, we will only visit the URL of the application, and later, we will create an interceptor to demonstrate how you can use the fixtures to provide mocked data:

```
beforeEach(() => {
  cy.visit('');
});
```

In the preceding code, we defined the `beforeEach()` function, and inside the callback, we visited the base URL of our application using the `visit()` method on the `cy` object. The `cy` object is a global helper object provided to you by the Cypress framework used for all sorts of things, such as visiting pages, accessing page and window objects, reacting to events, setting up interceptors, and waiting for requests. In the preceding example, we used the `visit()` method. We provided it with an empty string as a function parameter to indicate Cypress should visit the base URL of our application.

After defining the `beforeEach()` function and visiting your application base URL, let's add our first simple e2e test. As with unit tests, your test cases are defined using the `it()` function. Similar to the unit tests, your `it()` functions receive a description and a callback function. Inside the callback function, you write the code for your test cases. Let's start simple and write an e2e test to check whether the application redirects to the `expenses-overview` route when we visit the base URL:

```
it('should redirect to the expenses-overview page when we load the
root application route', () => {
  cy.url().should('<equal', 'http://localhost:4200/expenses-overview');
});
```

As you can see in the preceding code, we describe the test case and then write the logic for our test case inside the callback of the `it()` function. For the prior test case, we only need one line of logic. The `beforeEach()` function will open the application on the base URL. When we open our app on the base URL, we should be redirected to the `expenses-overview` route, so when we reach the `it()` function, the application should be redirected to the `expenses-overview` route. In this test case, you only have to assert whether the current URL equals `http://localhost:4200/expenses-overview`. In your Cypress tests, you generally get a page or window element and interact with them or assert the text, CSS classes, or attributes; in this case, you get the browser URL and assert whether the URL equals the text you expect.

The `cy` object exposes most window objects by default; if you want to access elements from within the HTML structure of your application, you can use the `.get()` method on the `cy` object. In our example, we are interested in the URL, which is located in the `location.href` property of the `window` object. The `cy` object exposes the `location.href` property by default using the `.URL()` method.

After you get the element you want to assert, you can chain the `.should()` assertion method to make the assertion you want to make. The `.should()` method takes in two parameters: an assertion type and a value with which to perform the assertion. In our case, we provided the `.should()` method with the `equal` assertion type and provided the `http://localhost:4200/expenses-overview` value to check whether our provided value equals the element we want to assert—in this instance, the URL.

You can find a list of all assertion types in the official Cypress documentation: `https://docs.cypress.io/guides/references/assertions`.

Now that we've written the first e2e tests and explained how everything works, let's run the e2e test and see whether the test will succeed. You can start your e2e tests by running the following terminal command at the root of your Nx monorepo:

```
nx e2e <project-name>
```

In the preceding command, you need to replace `<project-name>` with the name found in the `project.json` file of the e2e project you want to run. Just as with the unit tests, you can also run the e2e tests for multiple projects using the `run-many` command or for affected projects with the `affected` command.

When running one of the aforementioned commands, the e2e tests will run headlessly, meaning no browser will be opened to execute your e2e tests. Running your e2e tests headlessly is ideal if you want to run your tests in a build pipeline or another environment where you don't have access to a browser.

It is nice to see Cypress executing the tests in a real browser during the development process. When Cypress executes the tests in a real live browser, you can better understand why tests are failing. There is a nice user interface allowing you to easily spot what tests are failing and navigate to the specific steps in which they're failing. To run the e2e tests in a real live browser, you can use the following command:

```
nx e2e <project-name> --watch
```

So let's change the `<project-name>` placeholder with `finance-expenses-registration-e2e` and run the test we created:

```
nx e2e finance-expenses-registration-e2e --watch
```

As shown in *Figure 11.1*, the Cypress UI will be started and ask you to choose a browser when you run the preceding terminal command:

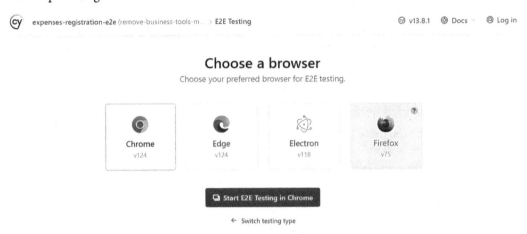

Figure 11.1: Cypress UI start screen

Select the **Chrome** browser and click on **Start e2e Testing in Chrome**, as shown in *Figure 11.1*. After starting the e2e tests in the Chrome browser, a new Chrome browser will be opened for you, showcasing all your e2e specs. The e2e specs are the `.cy.ts` files located inside the `e2e` folder of your e2e project; in our case, we only have one file, the `expenses-registration.cy.ts` file. When you click on the file name, Cypress will run the e2e tests for that specific file. In *Figure 11.2*, you can see what the test run looks like for our `expenses-registration.cy.ts` file:

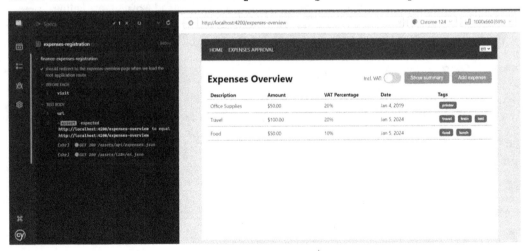

Figure 11.2: Cypress UI test run screen

As seen in *Figure 11.2*, on the left side, you can see the tests that are being executed and whether they pass or fail, and on the right side, you can see the application and what Cypress is doing inside the application.

To summarize, Nx creates an e2e project for each application you generate using the Nx cli or Nx console. You create .cy.ts files inside the e2e folder to define your test cases. Test cases are grouped using the describe() function and the beforeEach() function, which can be used to execute logic before each e2e test. The test cases themselves are defined using the it() function, and inside the callback of your it() function, you define the test logic. You can get elements using the cy object and assert values using the .should() method combined with an assertion type and a value to assert. When you define your tests, you can execute them in a real browser or headlessly if you want to run them inside an environment where you don't have access to a browser.

Now that you have learned the basics of e2e testing, created your first test, and run your test using the Cypress UI, let's add some extra e2e tests to learn about additional concepts and patterns commonly used within e2e testing.

Defining page objects for e2e testing

A common pattern in e2e testing is the **page object pattern**. When using the page object pattern, you abstract the selection of page elements away from the actual tests, resulting in more readable and maintainable tests. To demonstrate the page object pattern, let's first create a new e2e test without using the page object pattern and then adjust the new test by using the page object pattern. The new test case will check whether the **Show summary** button is shown by default, and when we click the button, the summary will be shown. Additionally, the test checks whether the button text is changed to **Hide summary** and whether the summary disappears if we click on the button again.

To create this test case, let's start by defining the it() function and provide it with a description for the test case:

```
it('should toggle the summary and adjust the button text', () => {});
```

Now that the it() function has been defined and we have provided a fitting description, we need to add the testing logic inside the callback of the it() function. First, we need to get the button we used to toggle the summary. As mentioned before, you can get elements from within your application using cy.get(). You provide the .get() method with a query selector; these selectors work identically to jQuery selectors:

```
cy.get('business-tools-monorepo-expenses-overview-page > div > div >
div > button:nth-child(2)');
```

In the preceding code snippet, you can see we use `cy.get()` and provide it with the query selector to get the **Show summary** button. If you're not familiar with jQuery selectors, you can alternatively copy the selector through the **DevTools** of the Chrome browser. Simply inspect the HTML page, find the element you want to use inside your Cypress test in your **DevTools**, right-click on the element, and select **Copy | Copy selector**. In *Figure 11.3*, you can see where you can copy the selector:

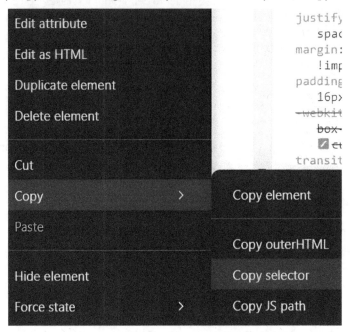

Figure 11.3: DevTools Copy selector

The selectors copied using the **DevTools** always start at the root of your HTML document and can be simplified by removing the beginning of the selector. Now that you know how to select elements so you can use them within your tests, let's write the rest of the test logic:

```
it('should toggle the summary and adjust the button text', () => {
    const button = () => cy.get('business-tools-monorepo-expenses-
overview-page > div > div > div > button:nth-child(2)');
    const summary = () => cy.get('table > tr.summary > td');
    button().should(<contain>, <Show summary>);
    summary().should(<not.exist');
    button().click();
    button().should(<contain>, <Hide summary>);
    summary().should(<exist>);
    button().click();
```

```
    button().should(<contain>, <Show summary>);
    summary().should(<not.exist');
});
```

In the preceding code, we first define two constants, one for `button` and one for `summary`. As you can see, we use a function that returns the `cy.get()` method. Directly assigning the return value of `cy.get()` to a variable is an anti-pattern because you can modify the return value during your tests. Commonly, when you get the button, you want to get it unmodified, so we create a function that returns the `cy.get()` function call and assign that to our variables.

After defining the two constants, we check whether the button contains the **Show summary** text and whether the summary element doesn't exist yet. Afterward, we click the button and check whether the button text is changed to **Hide summary** and whether the summary element exists. Lastly, we click the button again and see if everything is toggled back to its initial state.

If you run your e2e tests now, you'll find that the test succeeds. While there is nothing wrong with this test, there are some things we can do to clean it up a bit. First, we can simplify the selectors by adding a `data-test-id` attribute on the HTML elements we want to select. The `data-test-id` attribute is a simple HTML attribute commonly added to elements you want to use in your e2e tests. So, let's add the attribute to the button and summary element inside your `expenses-overview-page.component.html` file:

```
<button data-test-id="show-summary-btn"
(click)="onSummaryChange()">{{summaryBtnText()}}</button>
```

In the preceding code, you can see we've added the `data-test-id` attribute to the button and provided it with a value of `show-summary-btn`. Next, we will do the same for the element where we show the summary:

```
<td data-test-id="summary">Total: {{expensesVm.total}}</td>
```

Now you've added the `data-test-id` attributes, you can simplify the selectors used inside your e2e test. Instead of the long selector used to select the `button` and `summary` elements, you can use the following syntax:

```
cy.get('[data-test-id="show-summary-btn"]');
cy.get('[data-test-id="summary"]')
```

As you can see, this dramatically simplifies the selectors for your HTML elements. Besides simplifying the selectors by introducing the `data-test-id` attributes, you can create a function that checks the button text and checks whether the summary element exists. You don't have to repeat it three times inside your test.

Now that we have defined the logic for the test case, it's time to improve it and move some logic to the page object file. Start by removing the `app.po.ts` file from the support folder and adding a new `expenses-overview.po.ts` file. As you might have figured out, the `.po.ts` is short for `.page-object.ts`. Inside the `expenses-overview.po.ts` file, you will define all the logic to get the elements needed for the e2e tests for the `expenses-overview` page. By abstracting the element selection to this page object file, you can easily reuse them, making your e2e tests smaller and easier to read, write, and maintain. Currently, we only have two elements we can move to the page object file—the summary button and the summary element:

```
export const showHideSummaryBtn = () => cy.get('[data-test-id="show-summary-btn"]');
export const summaryValue = () => cy.get('[data-test-id="summary"]');
```

In the preceding code snippet, you can see that we moved the two constants defined inside the test case to the page object file and exported them so they can be accessed inside the e2e tests. We also gave the two constants a more descriptive name. Now, inside the `expenses-registration.cy.ts` file, import the two constants and adjust the e2e test to use the imported constants. If you now need the button or summary type in another e2e test, you can simply use the one defined inside the page object file instead of redefining the logic to get the element.

Inside the page object file, we can also add the function to check the `toggle summary button` text and the visibility of the summary itself:

```
export function summaryIsShwon(isShown: boolean) {
    showHideSummaryBtn().should('contain', isShown ? 'Hide summary' : 'Show summary');
    summaryValue().should(isShown ? 'exist' : 'not.exist');
}
```

After adding the preceding function, let's update the e2e test to use the function and see the final result after adding the page object file:

```
it('should toggle the summary and adjust the button text', () => {
    summaryIsShwon(false);
    showHideSummaryBtn().click();
    summaryIsShwon(true);
    showHideSummaryBtn().click();
    summaryIsShwon(false);
});
```

As you can see in the preceding code, now that we use the page object pattern, the test is much easier to understand and needs fewer lines of code; besides that, the code is easier to reuse in new test cases.

To summarize, you select elements using the `cy.get()` method combined with selectors identical to jQuery selectors. To simplify your selectors, you can use `data-test-id` attributes, and by using the page object pattern, you can abstract the element selection logic away from your test cases, making your tests easier to read, write, and maintain. Now that you have a better grasp of how to select elements and how the page object pattern can help you write better e2e tests, let's learn how to intercept requests and use mock data from your fixtures inside your e2e tests.

Using fixtures in your e2e tests

Fixtures are used to provide your e2e tests with specific mock data. Using mock data for your e2e tests ensures that you have stable data with which to run your e2e tests. Often, you need to run your e2e tests in an environment where you don't have access to an API or mocking service; in this case, you can use the data defined in your fixtures. Another common scenario is that you run your e2e tests in your test or acceptance environment, and the data on these environments isn't always stable and might change over time, resulting in failing tests. So, depending on your environment, fixtures can provide you with additional stability, ensuring that your tests don't fail based on the data but only fail if you actually break something within your application code.

Let's first run the e2e tests in production mode to demonstrate why you need fixtures. Our `mock.interceptor.ts` file will not return the mock data if we serve a production build of our application. You can use the following terminal command to run the e2e tests with a production build of the application:

```
nx e2e finance-expenses-registration-e2e --watch
--configuration=production
```

After running the preceding command, you'll notice the application doesn't have any data to display during the e2e tests. For our current test cases, this is no issue, but when you have more tests, this will most likely result in some failing tests. Instead of relying on `mock.interceptor.ts`, we can use the fixtures to provide data during the e2e tests.

To use the data from your fixtures, you first need to add a file with mock data inside the `fixtures` folder. We will use the same mock data as we use for our mock interceptor, so start by copying the `expenses.json` file inside the `assets/api` folder from your *expenses-registration application* to the `fixtures` folder of your e2e project.

After copying the `expenses.json` file, you need to adjust the `beforeEach()` function inside the `expenses-registration.cy.ts` file. Inside the `beforeEach()` function, you need to set up an interceptor to intercept the API request we make to get the expenses and provide it with a file from your fixtures:

```
beforeEach(() => {
  cy.intercept('GET', '**/api/expenses', { fixture: 'expenses.json'
  }).as('getExpenses');
  cy.visit('');
  cy.wait('@getExpenses');
});
```

In the preceding code, you can see we set up the interceptor using the `cy.intercept()` method. The `cy.intercept()` method first takes a string to define what type of API request you want to intercept; in our case, we want to intercept a GET request. Next, you need to provide the API URL you want to intercept, and lastly, you need to provide an object with a fixture property assigned with the fixture file you want to use as a response for the intercepted request. At the end of the `cy.intercept()` method, we chain the `.as()` method and provide that with an alias for the interceptor; in this case, we used `getExpenses`.

After setting up the interceptor, we define the `cy.visit()` method to visit the application page, just like we did before. After the `cy.visit()` method, we define the `cy.wait()` method, indicating that Cypress must wait for the interceptor we set up. Then, provide the `cy.wait()` method with the interceptor alias prefixed with an @ sign.

The preceding steps involve everything required to use the fixture file for your mock data during the e2e tests. If you need to set up the same interceptor and visit the same page for multiple testing files, you can abstract the logic away into a function and call that function inside the `beforeEach()` callback so that you don't have to repeat yourself multiple times.

You can test whether the interceptor and fixture work by running the e2e tests in production using the following command:

```
nx e2e finance-expenses-registration-e2e --watch
--configuration=production
```

After running the preceding command, you'll notice that the application shows data again when running the e2e tests.

To summarize, you learned that fixtures can be used to provide mock data during your e2e tests. Using mock data can provide your tests with additional stability and help you run them in an environment where you don't have access to an API or mocking service. You use fixtures by setting up an interceptor in the `beforeEach()` function of your tests and providing the interceptor with the fixture file.

Summary

In this chapter, you learned about automated application testing. You learned that unit tests are used to test small code units in isolation to ensure the code implantation works as expected. e2e testing tests applications from the user's perspective and checks whether the correct values are displayed and user interactions are processed and rendered correctly in the application view. Component testing is a relatively new concept, comparable to e2e testing, but instead of compiling and testing an entire application, component testing focuses on testing a single component from the perspective of a user. Lastly, integration tests are used to check how different modules and elements of your software integrate together. Integration tests can be implemented on various levels, for example, to check whether code implementation remains working when you combine multiple components and services or whether your application still works when you combine and deploy various Angular applications and backend APIs as a single product for your customers.

After learning about the different types of tests, you created your own unit and e2e tests. You learned about the `describe()`, `beforeEach()`, and `its()` functions and how they can be used for e2e and unit tests. You learned how to use mock data in e2e tests and stub components and services in your unit tests. Asserting values in unit tests is done using the `expect()` function, and in e2e tests, you use the `cy.should()` method.

Lastly, you learned about different terminal commands to run tests for individual files, single projects, multiple projects, or projects that are affected by your changes. In the next and final chapter of this book, you'll add the finishing touches and learn about the different steps you need to take to deploy Angular applications in your Nx monorepo.

12

Deploying Angular Applications

In this last chapter, you will deploy the demo application we created to GitHub Pages. You will learn to lint and build Angular applications and libraries inside your Nx monorepo. We will explore what Angular and Nx create for you when you build an Angular application and what it is you deploy to your hosting platform. When you know how to build your Angular applications, we will inspect the application build and analyze the different bundles inside our build to determine where we can make some improvements to reduce the size of our application bundles. After analyzing the application bundles, we will first host the production build of our application locally. Next, we will dive into GitHub Pages and GitHub Actions to create an automated deployment process. GitHub Pages is a static website hosting service that allows you to host your static websites and applications for free. GitHub actions are used to set up **continuous integration and continuous delivery (CI/CD)** flows to automate your linting, testing, and deployment process. The CI/CD process you create will automatically lint, test, and deploy your Angular app inside the Nx monorepo to GitHub Pages when you merge code to your main GitHub branch.

This chapter will cover the following topics:

- Building and linting Angular applications inside your Nx monorepo

- Analyzing your build output

- Automatically deploying Angular applications

Building and linting Angular applications inside your Nx monorepo

This section will explore the steps required to build and lint Angular applications within an Nx monorepo. We start by learning how to lint projects inside your Nx monorepo. Next, we will explore different configurations for the linting process and how to apply them. Once you know how and why to lint your projects before you create a build, we will learn about building projects inside the Nx monorepo.

Furthermore, we'll explore how Nx enhances the build process by leveraging advanced caching and parallel task execution to optimize build times. Nx's incremental builds significantly reduce the build times for large and complex projects, ensuring that only necessary parts of the application are rebuilt. The incremental builds and Nx cache will speed up your CI/CD pipelines and the overall deployment process. Let's start at the beginning and learn about linting projects inside your Nx monorepo.

Linting Nx projects

Linting is a crucial step in the software development process. It involves analyzing your code for potential errors, enforcing coding standards, and maintaining code quality. Linting helps catch issues early in the development cycle, reducing the likelihood of bugs and ensuring that your code base adheres to best practices.

Linting improves the maintainability and readability of your code and enhances collaboration within your team by providing a consistent coding style. We already talked about linting in *Chapter 1*. Depending on your setup, you will be notified about linting errors during your development process by ESLint. Still, it's important to always run the linting process in your CI/CD pipelines before your code is merged or deployed. Running the linting process inside your CI/CD pipelines ensures code that doesn't adhere to your configured guidelines is not merged. By integrating linting into your pipelines, you can ensure that your projects are robust, reliable, and ready for production, leading to more efficient and error-free builds.

As with all other tasks such as testing and building, you can perform linting using the Nx console or by running a terminal command.

Linting using the Nx console

If you want to lint a project using the Nx console, follow these steps:

1. Click on the Nx logo on the left side of VS Code.
2. Locate the project you wish to lint under the **PROJECTS** tab.
3. Hover over lint.

4.   Click the play button:

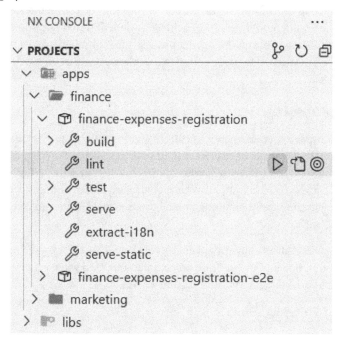

Figure 12.1: Nx console lint

Figure 12.1 shows a visual representation of the Nx console and where to find the lint task. When you click the play button for the lint task, a terminal will open inside VS Code, where you can see the result of the linting process. In the case of the *finance-expenses-registration application*, the linting process will pass without any error or warning regarding your linting rules. There will be a warning for Nx Cloud, but this can be ignored because we aren't using Nx Cloud. When using the Nx console, you can only lint your projects individually. Alternatively, with the Nx console, you can lint your projects using terminal commands.

Linting using the terminal command

You can lint a single project by running the following terminal command at the root of your Nx monorepo:

```
nx run <project name>:lint
```

In the preceding terminal command, you need to replace <project name> with the name of the project you want to lint. This project name can be found inside the project.json file of the project you want to target. So, to lint the *finance-expenses-registration application*, you run the following command:

```
nx run finance-expenses-registration:lint
```

Running the linting process for a second time was a lot faster due to the Nx cache. Inside the terminal, you'll find Nx read the output from the cache instead of running the command for 1 out of 1 tasks, indicating Nx took the linting output from the cache because it detected nothing changed since we last linted the project. When running the linting process for multiple projects, the Nx cache can significantly reduce the time needed for your linting to complete, greatly improving your development process and experience.

Linting multiple projects

In the preceding examples, we've seen how to run linting for a single project, but in most cases, especially inside your CI/CD pipelines, you want to lint multiple projects. When you run linting in a CI pipeline (a pipeline you run before each code merge), you commonly wish to lint projects that are affected by the changes you're about to merge; alternatively, when running a CD pipeline that deploys your system, you commonly run linting on all projects inside your monorepo or all projects related to the deployment.

To run linting for all projects affected by your changes, you can utilize the nx affected command. The terminal command is similar to the affected command we used when running tests for affected projects; you only change the target job from test to lint:

```
nx affected -t lint -base=main
```

As you can see in the preceding command, you use the affected keyword, add the -t flag, and define the target task you want to run. In this scenario, it is lint. After defining the task, we also need to configure the -base flag and provide it with the base branch to compare the changes with; commonly, this will be your main branch. Instead of manually defining the -base flag, you can also configure the defaultBase property inside your nx.json file. To configure your default base branch, add the following inside the root of your nx.json file:

```
"affected": {
   "defaultBase": "main"
}
```

After adding the preceding code, you can run the affected command without providing the -base flag. Currently, if you run the affected command, you'll notice all projects inside the monorepo are linted. All projects are linted because Nx doesn't have an affected lint cached for your defaultBase property. Besides all projects being linted, you will also notice that the linting process fails for 4 out of 13 projects.

Fixing linting errors

Before we continue, let's fix the linting errors so that all linted projects succeed.

The --fix flag

You can manually go through all the errors in the terminal output and fix them one by one, but most linting errors are easy to fix, and the Nx linting process provides a way to solve all easy linting errors for you. Simply append the `--fix` flag to your terminal command, and Nx will automatically fix the issues it can fix for you. To get the best result when using the `--fix` flag, you need to run the linting process for each project individually. So, let's take the projects with linting errors and run them one by one with the `--fix` flag appended to the command:

```
nx run workspace-generators-plugin:lint --fix
nx run shared-ui-common-components:lint --fix
nx run shared-data-access-generic-http:lint --fix
nx run shared-util-custom-decorators:lint --fix
```

After running the aforementioned commands, all your linting errors are fixed. You can now rerun the affected `lint` command, and the linting for all projects will succeed.

The fix option

As you might imagine, running the linting command for each project with linting errors can become cumbersome, especially when your monorepo grows. Alternatively, you can configure the `fix` option inside the `project.json` file of individual projects. Inside the `project.json` file, you'll find a `lint` section, and inside this section, you can add the `fix` option like this:

```
"lint": {
  ......
  "options": {
    "fix": true
  }
}
```

After adding the preceding configuration to each project's `project.json` file, you can simply run the `lint affected` command, and Nx will fix the linting issues it can for each project.

Other configurations

Besides the `--fix` flag and configuration, there are other useful configurations, such as allowing a maximum number of warnings before the linting process fails, outputting your lint results to a file, or passing the linting process even if there are linting errors. You can find all configurations in the official Nx documentation at the following URL: `https://nx.dev/nx-api/eslint/executors/lint`.

Targeting multiple projects when running a linting command

After looking into the remaining configuration options, I will briefly explain how you can target multiple projects simultaneously when running a linting command. You can use the `run-many` command to do this, similar to what we did in *Chapter 11* when testing multiple projects with a single command. You use the `run-many` command combined with the `-p` flag. The `run-many` command allows you to run tasks for many projects, whereas the `-p` flag will enable you to specify specific projects:

```
nx run-many -t lint -p project1 project2
```

In the preceding example, you perform linting for `project1` and `project2`. Alternatively, you can run the linting process for projects with a specific tag by using the following command:

```
nx run-many -t lint --projects=tag:type:ui
```

As you can see in the preceding command, instead of defining project names, we defined the tag we wanted to target.

Now that you know how to lint specific projects and how to configure the linting process, let's dive into the next topic: building your projects inside the Nx monorepo.

Building your Angular libraries and applications

To deploy your Angular projects, you need to create an application build. When you create an application build, the build process creates a bundle of code that your runtime can execute and run. In the case of Angular applications, the browser will be the runtime that interprets and runs the application code. So, to deploy the demo application we created, you first need to make an application build using the `nx build` command.

Behind the scenes of creating an application build

Before running the `nx build` command for our *expenses-registration application*, let's examine in detail what happens when we run a `build` command:

1. **Preparation phase**

    *   **Build configuration**: Nx reads the build configurations from the `project.json` files of the projects you're about to build. This is similar to the standard Angular CLI but with support for multiple projects in a monorepo, and Nx analyzes the `project.json` files instead of the `angular.json` file in a regular Angular project.

    *   **File replacements**: Nx checks if there are file replacements configured inside the `project.json` file. If you configured file replacements for the build targets, Nx will replace the files before continuing the build process.

2. **Compilation phase**

    - **TypeScript compilation**: Nx uses the Angular Compiler, which is called `ngc`, to transpile TypeScript code into JavaScript, handling Angular decorators and templates.

    - **Ahead-of-time (AOT) compilation**: As with the Angular CLI, Nx performs AOT compilation for production builds (unless configured differently), converting Angular templates and components into efficient JavaScript code that can be executed by the browser without the need to compile the code before it can be rendered.

3. **Bundling phase**

    - **Module resolution**: Nx uses the configured build tool for module resolution and creating a dependency graph, starting from the entry point(s).

    - **Tree shaking**: Nx performs tree shaking to remove unused code, reducing the final bundle size.

    - **Code splitting**: Nx splits different parts of your application into multiple JavaScript bundles that can be lazy loaded by the browser.

    - **Asset optimization**: Nx optimizes CSS, HTML, images, and other assets, similar to the Angular CLI.

4. **Minification and uglification**

    - **JavaScript minification**: Nx minifies JavaScript code to reduce file size.

    - **Uglification**: Nx further obfuscates the JavaScript code.

5. **Hashing and cache busting**

    - **File hashing**: Nx appends hashes to filenames of generated bundles and assets for cache busting.

6. **Generating HTML**

    - **Index HTML generation**: Nx generates or updates the `index.html` file, including references to the hashed JavaScript and CSS bundles.

7. **Deployment artifacts**

    - **Output directory**: Nx places the final build artifacts in the `dist` folder or some other configured output path.

The preceding steps resemble the regular build flow when using the Angular CLI. Yet, Nx has some additional features such as affected builds, the `run-many` command, incremental builds, and caching, making it easier to create application builds for multiple projects inside a monorepo and speeding up build times. Now that you have a better grasp of what is happening when building your projects, let's create an application build for the expense registration application.

> **Important notice!**
>
> The *expenses-registration application* we made is just a simple demo application and not really meant for production purposes. There are still plenty of improvements to make, pages to add, and code to clean up before you actually deploy the application. For demonstration purposes, we will be deploying the application to GitHub Pages, but it's not meant to be used for anything other than demonstration purposes.

Running the build command

You can run the `build` command for the *expenses-registration application* in the terminal or Nx console. As shown in *Figure 12.2*, you can find the `build` command in the Nx console above the `lint` command:

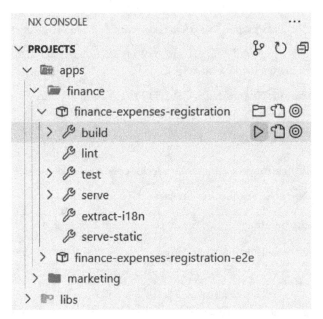

Figure 12.2: Nx console build

To run the `build` command, click the play button next to the `build` keyword in the Nx console. Nx configures the production build as your default; if you want to target another build configuration, you can expand `build` inside the Nx console to see all build configurations you have. You can then run a specific build configuration by clicking on the play button next to the name of the build configuration you want to run. For each project, Nx creates a production and development build configuration for you inside the `project.json` files. When needed, you can create additional build configurations; for example, when you have multiple staging environments such as test and acceptance that require a deviation from the production or development build.

The Nx console is an extension for your code editor, and as such, it isn't available inside your CI/CD pipelines. Because the Nx console isn't available, you need to use the Nx CLI inside your pipelines to create application builds and perform other tasks such as building and testing your applications. The syntax for the `nx build` command is similar to the `test` and `lint` commands we used earlier; you only change the build target from `lint` or `test` to `build`:

```
nx build finance-expenses-registration
```

As you can see in the aforementioned command, we type `nx` to target the Nx CLI, followed by the task we want to run, and lastly, add the project name of the project we want to target. Nx will use the default build configuration, which is production unless you change it. If you want to target a different build configuration, you can add the `--configuration` flag to the `build` command:

```
--configuration=development
```

By adding the preceding flag at the end of your CLI command, you'll use the `development` build configuration instead of the production configuration. Alternatively, you can use the following syntax:

```
nx run finance-expenses-registration:build:production
```

The preceding command runs the production build for the `finance-expenses-registration` project. As we've seen with the `lint` and `test` commands, you can also use `affected` and `run-many` to build multiple projects or projects affected by your changes. Now that you know how to run `build` commands using the Nx console and the Nx CLI, go ahead and run the production build for the `finance-expenses-registration` project.

Fixing failing builds

After running the production build for the `finance-expenses-registration` project, you'll notice Nx tries to build the `finance-expenses-registration` project and all projects it depends on – in our case, three other libraries (`shared-ui-common-components`, `finance-data-access-expenses`, and `shared-util-form-validator`) – in our Nx monorepo. You'll also notice that three out of four builds fail. The three failing builds are the three libraries, and they all have the same error: `'updateBuildableProjectDepsInPackageJson' is not found in schema`.

To fix your builds, you need to remove the following configuration inside the `project.json` files of your libraries:

```
"updateBuildableProjectDepsInPackageJson": true
```

After removing the preceding configuration inside your `project.json` files, you can run the `build` command for the `finance-expenses-registration` project again, and your build process will now succeed. After the build is finished, you'll notice a `dist` folder has been created in the root of your Nx monorepo. The build output for the `finance-expenses-registration` project and all dependent buildable projects are located inside this `dist` folder. The build output for the `finance-expenses-registration` project can be found at this path: `dist/apps/finance/expenses-registration`.

When you deploy your application, you upload the output inside `dist/apps/finance/expenses-registration` to your hosting service. In the case of an Angular application, you can use a static hosting service such as GitHub Pages, Azure Blob Storage, Amazon **Simple Storage Service (S3)**, or any other static website hosting service you prefer. If you are using Angular Universal and **server-side rendering (SSR)**, you need a different hosting service, such as Azure App Service, but we will not cover that in this book.

In the last section of this chapter, we will use GitHub Actions to automatically deploy and host our demo application on GitHub Pages, but for now, let's see how we can serve our production build locally on our machine using `http-server`.

Locally serving the production build

To serve our production build, we need a server that can host our static files. There are many services out there that can do this, but we will be using `http-server`. To start, you need to install `http-server` on your machine using the following command:

```
npm install http-server -g
```

After installing `http-server`, you can use it to host the production build of your Angular application locally. If you don't have staging environments such as test and acceptance, using something such as `http-server` allows you to test if your build works as expected before deploying it to your production environment; still, in a professional setting, I would always recommend using a testing and acceptance environment where you can test your application in an environment that mimics the production environment.

To locally host your application using `http-server`, you need to run the following terminal command at the root of your Nx monorepo:

```
http-server dist\apps\finance\expenses-registration
```

After running the preceding command, a server will be created, hosting the static files inside the `dist\` `apps\finance\expenses-registration` folder. Inside the terminal, you'll find two URLs starting with your private IP addresses and ending with the `8080` port. You can visit either of these URLs inside your browser to see your application. When you visit the application in your browser, you'll see the requests that are made by your application inside the terminal. You may notice that the `api/expenses` request is failing; the reason for the failing API request is that we don't have a running API, and the mock API interceptor is disabled in a production build.

Because we disabled the mock data interceptor and we don't have a running API, you don't see any data inside the application when you visit it in the browser. For testing purposes, you can enable the mock interceptor for your production build by removing the following inside the `mock.interceptor.` `ts` file:

```
!isDevMode() ||
```

After removing the preceding code, the mock interceptor will also work for your production build. Create a new application build, serve the build using `http-server`, and revisit the application inside your browser (you probably need to remove the browser cache by opening the developer tools, right-clicking on the reload symbol in the browser, and selecting **Empty Cache and Hard Reload**). Now, when you visit the application on the URL listed inside the terminal, you should see the mock data, and the `api/expenses` request succeeds. Don't forget to revert the changes inside the `mock.` `interceptor.ts` file, as this is just for testing purposes; in a real production application, you want to use a real API.

So, to summarize, you learned how to lint projects inside your Nx monorepo and how you can automatically fix basic lint errors by using the `--fix` flag. You learned what happens behind the scenes when you create an application build for your Angular applications. You also learned how to run the `build` command for specific build configurations and where to find the output of your application build. Lastly, you hosted the build output locally using the `http-server` npm package. In the next section, you will learn how to analyze your application build so that you can easily identify which sections of your build can be worked on to reduce your bundle sizes.

Analyzing your build output

Inside your build output, you'll find different JavaScript files. Some of these files are your application bundles, which the browser will load to render the application for the end user. The size of these bundles directly impacts the performance of your application. JavaScript is slow to load for a browser, so the bigger your bundle size, the longer it takes for the browser to download the files, render something on the screen, and make the web page respond so that the user can interact with it.

To reduce your bundle sizes, you need an effective way to analyze your bundles. If you run the `build` command for the `finance-expenses-registration` project, you'll see a small report inside the terminal

The build report inside your terminal includes a list of the created bundles and the size of each bundle. While this list gives you some indication of the size of your bundles, you don't see what the bundles are made up of and where you can make some improvements to reduce the bundle sizes.

There is a tool named **Webpack Bundle Analyzer** that can give you a more detailed overview of your bundles. To use the Webpack bundle analyzer, you start by globally installing npm packages with the following command:

```
npm i -g webpack-bundle-analyzer
```

After installing the package globally, you need to install it as a dev dependency inside your Nx monorepo. You add the Webpack Bundle Analyzer tool as a dev dependency by running the following command in the root of your Nx monorepo:

```
npm i webpack-bundle-analyzer -save-dev
```

After running the preceding npm commands, you can start using Webpack Bundle Analyzer. The first step is creating a new production build for your application and including a `--stats-json` flag with the `build` command:

```
nx build finance-expenses-registration --stats-json
```

By adding the `--stats-json` flag, Webpack Bundle Analyzer will create a `stats.json` file inside the build output. The `stats.json` file is located inside the `dist\apps\finance\expenses-registration` folder, next to the rest of your application build. To inspect the `stats.json` file, you can run the following command at the root of your Nx monorepo:

```
webpack-bundle-analyzer dist/apps/finance/expenses-registration/stats.
json
```

Running the preceding command will give you a visual and detailed overview of your application bundles. The overview is opened in the browser and should look like this:

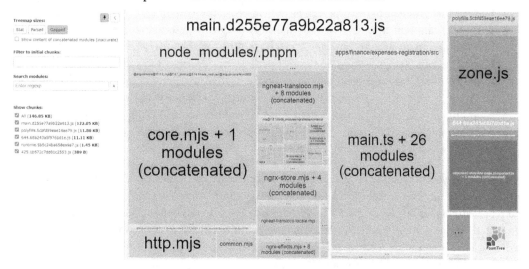

Figure 12.3: Webpack Bundle Analyzer

As shown in *Figure 12.3*, the Webpack Bundle Analyzer tool gives you a detailed and easy-to-read overview of your bundles and what is in each bundle. You can easily identify the large parts of each bundle so that you can see where you can make improvements. Because we have a simple application, there isn't much to adjust, but you can, for instance, see that the `ngrx-store` and `transloco` bundles are one of the largest elements we've added to our project. If your bundle sizes are becoming too large, you can try to find replacements for large npm packages you import, only import small sub-modules if they are available, or move code and other resources to lazy-loaded modules so that the browser can first render the page and lazy load additional resources.

As you can see in *Figure 12.3*, in the sidebar of Webpack Bundle Analyzer, you can control which bundles you show on the screen; this makes it easier to focus on specific bundles in your build output. Additionally, you can see the bundle sizes in three different formats – the **Stat**, **Parsed**, and **Gzipped** formats:

- **Stat** is the largest; this indicates the raw size of your bundle without any compressions or optimizations.

- **Parsed** is the same size as shown in the build output report of your terminal. This is the size of your bundles after some optimizations and compressions are applied.

- **Gzipped** is the smaller size. It would be best always to aim to Gzip your static build assets. As you can see, the Gzipped content is almost 69% smaller than that of the parsed assets. Having bundles that are 69% smaller will significantly improve the load times of your application. Most static application hosting services automatically Gzip your resources for you and provide the necessary configuration to unwrap Gzipped resources in the browser. You can check if your hosting service Gzipped your bundles by opening the developer tools in your browser, going to the **Network** tab, and checking the response header on the requests for your bundles. If you find the `Content-Encoding: Gzip` header, you know your hosting service enabled Gzipping for you; if the header is not included, you need to add Gzipping yourself. How to add Gzipping yourself falls out of the scope of this book. GitHub Pages will Gzip your Angular application for you.

To summarize, you now know how to analyze your application bundles effectively using Webpack Bundle Analyzer. If bundles are too large, you can try to move code to lazy-loaded modules and find replacements for large npm packages you import. You also learned that Gzipping your content significantly reduces your bundle sizes and that most modern hosting services handle Gzipping and the necessary configuration for you. In this book's next and last section, we will use GitHub Actions to create a CI/CD process that automatically deploys our demo application to GitHub Pages.

Automatically deploying Angular applications

In this last section, you will learn how to automatically deploy Angular applications inside your Nx monorepo to GitHub Pages. You will set up a CI/CD pipeline using GitHub Actions that deploys the demo application we've created whenever you merge code to your main branch on GitHub. We will create a `.yml` file containing all the steps necessary for the deployment. Inside the `.yml` file, we will use the YAML language, which is commonly used to create configuration files for various DevOps tools and programs.

Creating an access token

Before you start creating a `.yml` file, you need to create an access token inside your GitHub account. GitHub needs to know you're authenticated to deploy applications to GitHub Pages before it can perform the deployment. You provide the necessary authentication to GitHub, providing an access token to the deployment step inside your `.yml` file. So, the first thing you need to do is create an access token inside your GitHub account. Follow these steps:

1. Start by clicking on your account icon in GitHub and select **Settings**.

2. On the profile settings page, you need to scroll down and select **Develop settings** on the left side of the screen.

3. Now, again on the left side of the screen, click on **Personal access tokens** and then select **Tokens (classic)** in the drop-down menu.

4. Next, you need to click on the **Generate new token** button, and in the drop-down menu, you select **Generate new token (classic)**.

After following *steps 1* to *4*, you should land on the page shown in *Figure 12.4*:

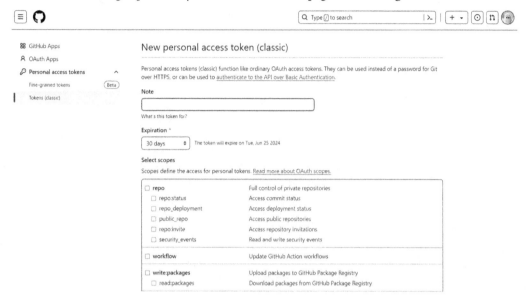

Figure 12.4: GitHub personal access token

5. Now, on the **New personal access token (classic)** page, as shown in *Figure 12.4*, enter GH_PAGES under the **Note** field. You can add any name here you want, but to follow along, you should use the GH_PAGES name.

6. For the **Expiration** field, select **No expiration**. In a real production environment, you might want an expiration date for security reasons, but this means you need to renew your token for the set expiration interval.

7. Under the **Select scopes** field, check all checkboxes up to gist.

8. After selecting the checkboxes under **Select scopes**, you can scroll down and select **Generate token**. After clicking on **Generate token**, you will see the key of your personal access token; it will look similar to this:

```
ghp_vnjAH0PRGRcIO6UQGO2GavdOUVUDPJ1Jrv77
```

Copy the key to your access token and store it safely; you will need it in the next step of your automated deployment configuration.

9. After copying and saving your key, go to the GitHub repository where you stored the code you created during the course of this book.

10. Next, you need to navigate to the repository settings by clicking on **Settings** in your navigation items, as shown in *Figure 12.5*:

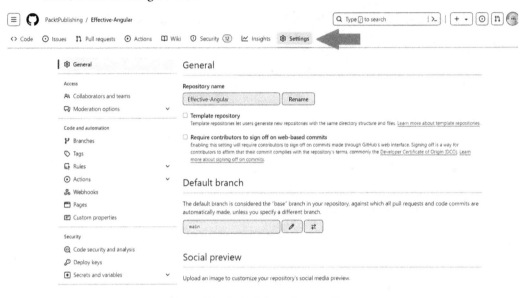

Figure 12.5: GitHub repository settings

11. Now, on the left side of the screen under **Security**, click on **Secrets and variables**, and in the dropdown, click on **Actions**.

This will bring you to the **Actions secrets and variables** page, as shown in *Figure 12.6*:

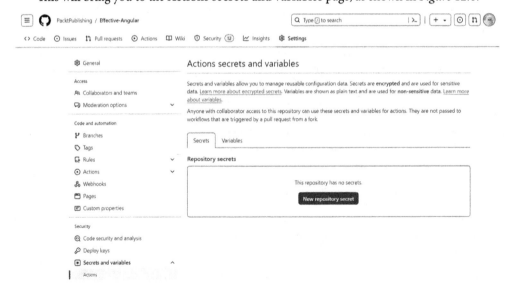

Figure 12.6: Actions secrets and variables

12. On the **Actions secrets and variables** page, you need to click on the **New repository secret** button.

13. Next, you'll be asked to provide a name and secret. Inside the **Name** field, provide GH_PAGES again, and in the **Secret** field, paste the key you saved from the personal access token.

14. After entering the necessary values, click on the **Add secret** button to save your repository secret.

If successful, you should now see your GH_PAGES secret under the **Repository secrets** section on the **Actions secrets and variables** page.

Creating a .yml file

Now that you've created a personal access token and a repository secret, we can continue by creating a .yml file.

To create a .yml file, follow these steps:

1. Start by clicking on **Actions** inside your repository navigation.

2. Next, click on **Set up a workflow yourself**.

   This will bring you to a page as shown in *Figure 12.7*:

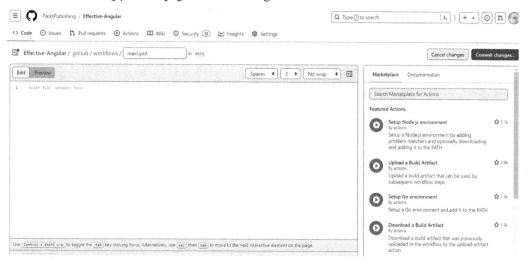

Figure 12.7: main.yml

As shown in *Figure 12.7*, GitHub created an empty main.yml file for you inside the .github/ workflows folder. You can directly edit the main.yml file in GitHub, alternatively, you can commit the file and edit it inside VS Code. For now, we will simply edit the file inside GitHub and start to add the necessary steps to deploy our demo application to GitHub Pages.

Adding a trigger

You start by adding a trigger to the `.yml` file. In our case, we want our `.yml` file to be triggered when we push changes to the main branch of our GitHub repository. You add this trigger by adding the following content to the `.yml` file:

```
# Add trigger
name: Build and Deploy Script
on:
  push:
    branches:
      - main
```

The preceding code snippet starts with a comment; next, we give a name to the step we're declaring, then we use the `on:` keyword to indicate we're defining a trigger, and underneath it, we define the configuration of our trigger – in this case, on push of the main branch. The indentation in `.yml` files is important, so make sure each line has an extra tab compared to the line above it.

Defining jobs to run and the related steps

After defining the trigger, you need to define jobs to run and the steps for the jobs you want to run.

In our case, we only have one job, which is a `build` job consisting of multiple steps. Start by defining `jobs` and `build` keywords, followed by defining the OS for the VM the build job will run on:

```
# Jobs to run
jobs:
  build:
    runs-on: ubuntu-latest
```

Underneath the `runs-on` configuration of the build job, we will define the steps needed to perform the build job. The first step you need to perform is to check out the GitHub repository:

```
    # Checkout repository
    - name: Checkout Repository
      uses: actions/checkout@v4
      with:
        fetch-depth: 0
```

As you can see in the preceding snippet, the indentation for this step starts at the same place as the `runs-on` configuration. We start by defining a comment; next, we define a name for the step and what action the step uses. Lastly, we provide additional configurations for the step – in this case, `fetch-depth: 0`, which indicates the fetch history of all branches and tags inside your GitHub repository.

After fetching the GitHub repository, you need to install NodeJS on the Ubuntu machine with the following step:

```
# Install NodeJS
- name: Adding Node.js
  uses: actions/setup-node@v2
  with:
    node-version: 20.1.0
```

After installing NodeJS, you need to install `node_modules`. To speed up the build job, we will use pnpm instead of npm. Simply add the following step to your `.yml` file to install `node_modules` for your Nx monorepo:

```
# Setup pnpm
- uses: pnpm/action-setup@v2
  with:
    version: 8.14.1
- run: pnpm install --frozen-lockfile
```

The next step in our process will be linting all affected projects:

```
# Lint affected projects
- name: Lint affected
  run: pnpm nx run-many -t lint --base=main --no-cloud
```

As you can see, we added a new flag to the command we haven't seen before. We've added the `--no-cloud` flag to indicate we're not using Nx Cloud; otherwise, your pipeline will fail by trying to connect with Nx Cloud. Next, we will add an extra step to unit test our application:

```
# Unit testing finance-expenses-registration
- name: Unit test finance-expenses-registration
  run: pnpm nx run finance-expenses-registration:test —no-cloud
```

For simplicity, we will skip **end-to-end** (**e2e**) tests and only run unit tests for the `finance-expenses-registration` project. If you already fixed all unit tests within the monorepo, you can change the `.yml` file to run tests for all affected projects. After running the `lint` and `test` commands, we will build the `finance-expenses-registration` project with the following step:

```
# Build finance-expenses-registration application
- name: Build Angular App
  run: pnpm nx build finance-expenses-registration --no-cloud
--base-href /Effective-Angular/
```

As you may have noticed, we added the `--base-href` flag to the `build` command. We need to add a base `href` attribute because GitHub will deploy your application to `<GitHub account name>.github.io/<repository name>`. If you don't use `Effective-Angular` as your repository name, you need to change the value of the `--base-href` flag to your own repository name.

After building the repository, you need to initialize Git on the VM so that you can deploy it to GitHub Pages. Simply add the following step to your `.yml` file:

```
# Setup Git on the VM
- name: Set up Git
  run: |
    git config --global user.email "youremail@gmail.com"
    git config --global user.name "your-git-username"
```

After adding the preceding step, we can move to the last step inside our build job, this being the deployment of our application to GitHub Pages.

Deploying the application to GitHub Pages

We will be using an npm package named `angular-cli-ghpages` to do the deployment of the application. Add the following step to your `.yml` file:

```
# Deploy the finance-expenses-registration to GitHub pages
- name: Deploy to gh pages
  run: |
    npx angular-cli-ghpages --dir=dist/apps/finance/expenses-
registration
  env:
    CI: true
    GH_TOKEN: ${{ secrets.GH_PAGES }}
```

As you can see in the preceding code, we run a terminal command using `angular-cli-ghpages` to deploy the application, and we need to provide the GH_PAGES token we created earlier so that GitHub knows we have the necessary authorization to do the deployment. After adding the deployment step, you've added all steps needed to automatically deploy your demo application to GitHub Pages.

You can now click on the **Commit changes...** button, and in the modal that pops up, click on **Commit changes**. After committing the `.yml` file to your main branch, the `.yml` file will immediately be triggered. You can see this by navigating back to the **Actions** tab of your repository, and there you should see your running workflow. Unfortunately, the workflow currently fails.

Fixing workflow fails

The GitHub Actions workflow failed because of multiple reasons, so let's fix the problems one by one. First, you need to install the `angular-cli-ghpages` packages inside your Nx monorepo. Simply run the following command to install the package in your monorepo:

```
npm i angular-cli-ghpages
```

Next, we use pnpm instead of npm in our GitHub Actions workflow. Because of this, we need to commit a `pnpm-lock.yaml` file to our GitHub repository. To generate this `pnpm-lock.yaml` file, start by globally installing pnpm with the following npm command:

```
npm install -g pnpm
```

After installing pnpm, you can start using pnpm instead of npm for all npm commands you normally run. I would advise using pnpm as it's a lot faster compared to npm. Now, to generate your `pnpm-lock.yaml` file, simply run the following command:

```
pnpm install
```

The aforementioned command is similar to the `npm install` command, only using pnpm instead of npm. After running the `pnpm install` command, you can commit your changes and merge them with your main branch on GitHub to trigger the workflow again. As shown in *Figure 12.8*, under the **Actions** tab of your repository, you should now see two workflow runs, and the second run should succeed:

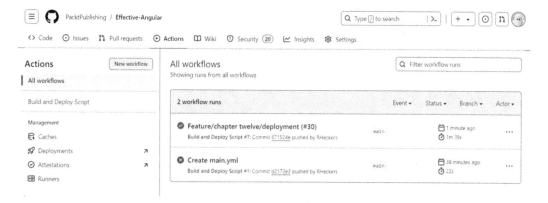

Figure 12.8: GitHub Actions workflow runs

You now have a successful build, yet you still need to handle the deployment step. Go to your repository settings, and on the left side, click on **Pages** and add the configuration as shown in *Figure 12.9*:

Build and deployment

Source

Deploy from a branch ▾

Branch
Your GitHub Pages site is currently being built from the gh-pages branch. Learn more about configuring the publishing source for your site.

🌿 gh-pages ▾   📁 / (root) ▾   Save

Learn how to add a Jekyll theme to your site.

Figure 12.9: Build and deployment

After configuring the branch to gh-pages, as shown in *Figure 12.9*, the deployment process will automatically start, and after a couple of minutes, you can find the URL of your application at the top of this GitHub page. Each time your workflow runs successfully, the deployment will now be done automatically. When you visit the demo application on the deployed URL, you'll notice nothing but your navbar is shown, because the application is deployed on the <GitHub account name>.github.io/<repository name> URL, and we need to account for the repository name when fetching the language files inside our application, so there is one last fix to be made.

Go to the app.config.ts file of the finance-expenses-registration project, and add the following provider at the top of your providers array:

```
{ provide: APP_BASE_HREF, useValue: isDevMode() ? '' : '/Effective-
Angular/' },
```

Next, go to your transloco-loader.ts file and adjust it to contain the following content:

```
@Injectable({ providedIn: 'root' })
export class TranslocoHttpLoader implements TranslocoLoader {
  private http = inject(HttpClient);
  protected readonly baseHref = inject(APP_BASE_HREF);

  getTranslation(lang: 'en' | 'nl') {
    return this.http.get<Translation>(`${this.baseHref}assets/
i18n/${lang}.json`);
  }
}
```

As you can see, we injected the `APP_BASE_HREF` injectable into the file and included the `baseHref` property into the URL used to fetch the translations. You can now push the changes and merge them with your main branch to trigger the workflow again. After your workflow run succeeds, you can click on **Deployments**, as shown in *Figure 12.10*:

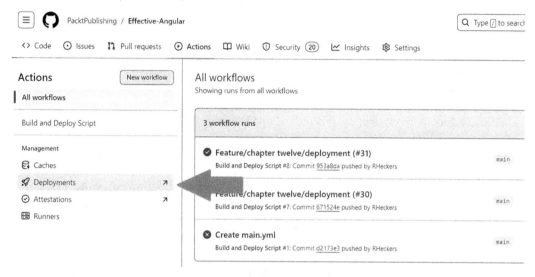

Figure 12.10: Deployments

When you click on **Deployments**, as shown in *Figure 12.10*, you'll be navigated to a page where you can see all deployments of your application. On this same page, you'll also find the URL of your deployed application and when it was last updated:

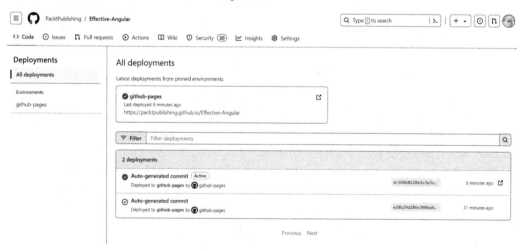

Figure 12.11: Deployments list

When your new deployment is ready, as shown in *Figure 12.11*, you can navigate to your deployed application in the browser again, and now you should see a working application, including a table and translated values. There is no data inside the table because the mock interceptor is disabled for your production build, and there is no API your application can connect with. You now have everything in place, and your application is linted, tested, built, and deployed automatically each time you push changes to your main GitHub repository.

As mentioned before, this is just a demo application, and there are still plenty of improvements to make. So, if you want, you can add more pages, create separate library components for the table and other features you want to have, and create and convert the application into something truly suited for production purposes. You now have all the building blocks needed to effectively use Nx and the Angular framework to develop robust and scalable applications.

Summary

We've reached the end of this book, and first of all, I want to thank you for reading the book and congratulate you on finishing it! You learned and did a lot during the course of this book. You created your own Nx monorepo, ready to scale to hundreds of Angular applications and libraries. You know how to configure the Nx monorepo to your personal needs and add the first libraries and applications inside your monorepo. After creating the monorepo and adding projects, you learned about built-in and custom Nx generators and how to use them to generate boilerplate code and bring consistency into your monorepo.

Next, we moved on to Angular and learned about the latest developments of the framework, as well as how to utilize powerful features such as component communication, Angular routes, and **dependency injection** (**DI**). You learned about pipes, directives, and animation after exploring the framework's newest and most powerful features. You saw how to create custom pipes and directives, learned about directive composition, and created reusable animation to make your application more appealing to the end user.

Further, we did a deep dive into the Angular `forms` module, learning about template-driven, reactive, and dynamic forms. Among other things, you learned about form validations, the form builder, error handling and creating dynamic forms, and form fields. After finishing with Angular forms, we returned to theoretical knowledge, exploring different best practices, conventions, and design patterns commonly used within Angular applications. You learned about some design patterns and best practices, setting you up for the next section of this book, which focused on reactive programming and state management. You also learned how to utilize RxJS and Signals to create reactive code. You learned about Observable streams, pipeable operators, and handling nested data streams. We converted code from RxJS to Signals to demonstrate the difference between the two and explained when to use which tool to reach the desired outcome. We also explored combining RxJS and Signals to get the best of both worlds. When we finished with the basics of RxJS and Signals, we used both tools to create a custom state management solution and connected it with the component layer of our application using a facade service. To finish our state management journey, we converted our custom state management solution to an implementation using NgRx, the Angular ecosystem's most powerful and popular state management solution.

When we finished with reactive programming and state management, we moved on to the last section of this book. You learned how to develop applications accessible to people from all over the world and of all abilities. You used `transloco` to add localization and internationalization to your applications so that people of different countries and languages can use your application with their preferred formats and languages. Next, you created unit and e2e tests using Jest and Cypress.

Lastly, we finished the book by creating an automated deployment process that uses GitHub Actions to deploy your application to GitHub Pages. After finishing this book, you will have all the knowledge needed to develop robust and scalable Angular applications using Nx and the latest Angular techniques. You can continue with the demo application we created by adding more pages, abstracting the table into its own component, creating a feature component for the expenses registration so that the page only has to declare one component, adding new features such as grouping expenses, planning expenses, uploading receipts, and whatever you seem fit. You can also start creating multiple applications within the Nx monorepo or start from scratch with your own project. Whatever you want to develop, you now have the tools and knowledge to effectively use Angular for applications of any scale.

Index

`packtpub.com`

Subscribe to our online digital library for full access to over 7,000 books and videos, as well as industry leading tools to help you plan your personal development and advance your career. For more information, please visit our website.

Why subscribe?

- Spend less time learning and more time coding with practical eBooks and Videos from over 4,000 industry professionals

- Improve your learning with Skill Plans built especially for you

- Get a free eBook or video every month

- Fully searchable for easy access to vital information

- Copy and paste, print, and bookmark content

Did you know that Packt offers eBook versions of every book published, with PDF and ePub files available? You can upgrade to the eBook version at `packtpub.com` and as a print book customer, you are entitled to a discount on the eBook copy. Get in touch with us at `customercare@packtpub.com` for more details.

At `www.packtpub.com`, you can also read a collection of free technical articles, sign up for a range of free newsletters, and receive exclusive discounts and offers on Packt books and eBooks.

Other Books You May Enjoy

If you enjoyed this book, you may be interested in these other books by Packt:

Angular Design Patterns and Best Practices

Alvaro Camillo Neto

ISBN: 978-1-83763-197-1

- Discover effective strategies for organizing your Angular project for enhanced efficiency
- Harness the power of TypeScript to boost productivity and the overall quality of your Angular project
- Implement proven design patterns to streamline the structure and communication between components
- Simplify complex applications by integrating micro frontend and standalone components
- Optimize the deployment process for top-notch application performance
- Leverage Angular signals and standalone components to create performant applications

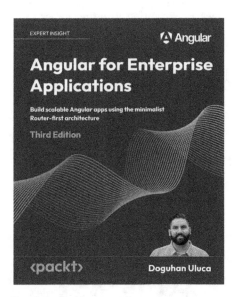

Angular for Enterprise Applications

Doguhan Uluca

ISBN: 978-1-80512-712-3

- Best practices for architecting and leading enterprise projects
- Minimalist, value-first approach to delivering web apps
- How standalone components, services, providers, modules, lazy loading, and directives work in Angular
- Manage your app's data reactivity using Signals or RxJS
- State management for your Angular apps with NgRx
- Angular ecosystem to build and deliver enterprise applications
- Automated testing and CI/CD to deliver high quality apps
- Authentication and authorization
- Building role-based access control with REST and GraphQL

Packt is searching for authors like you

If you're interested in becoming an author for Packt, please visit `authors.packtpub.com` and apply today. We have worked with thousands of developers and tech professionals, just like you, to help them share their insight with the global tech community. You can make a general application, apply for a specific hot topic that we are recruiting an author for, or submit your own idea.

Share Your Thoughts

Now you've finished *Effective Angular*, we'd love to hear your thoughts! Scan the QR code below to go straight to the Amazon review page for this book and share your feedback or leave a review on the site that you purchased it from.

`https://packt.link/r/1-805-12553-2`

Your review is important to us and the tech community and will help us make sure we're delivering excellent quality content.

Download a free PDF copy of this book

Thanks for purchasing this book!

Do you like to read on the go but are unable to carry your print books everywhere?

Is your eBook purchase not compatible with the device of your choice?

Don't worry, now with every Packt book you get a DRM-free PDF version of that book at no cost.

Read anywhere, any place, on any device. Search, copy, and paste code from your favorite technical books directly into your application.

The perks don't stop there, you can get exclusive access to discounts, newsletters, and great free content in your inbox daily

Follow these simple steps to get the benefits:

1. Scan the QR code or visit the link below

https://packt.link/free-ebook/978-1-80512-553-2

2. Submit your proof of purchase
3. That's it! We'll send your free PDF and other benefits to your email directly

www.ingramcontent.com/pod-product-compliance
Lightning Source LLC
LaVergne TN
LVHW081512050326
832903LV00025B/1462